DON'T DIE IN AUTUMN

The Magic & Madness of a Life for the Birds

DON'T DIE IN AUTUMN

The Magic & Madness of a Life for the Birds

ERIC DEMPSEY

Gill & Macmillan

Gill & Macmillan
Hume Avenue, Park West, Dublin 12
www.gillmacmillanbooks.ie

© Eric Dempsey 2015
978 0 7171 6579 7

Design by Make Communication
Print origination by Síofra Murphy
Printed and bound by ScandBook AB, Sweden

This book is typeset in Minion 12.5/15.5 pt.

The paper used in this book comes from the wood pulp of
managed forests. For every tree felled, at least one tree is
planted, thereby renewing natural resources.

A CIP catalogue record for this book is available from the
British Library.

5 4 3 2 1

For Ann and Tom Dempsey, my wonderful parents.

Also for Michael O'Clery, a true friend in every sense of the word.

And for Hazel, my best friend, my soul friend.

CONTENTS

ACKNOWLEDGMENTS

Sitting down to write this section is almost as challenging as writing a whole book. Where do I start?

I have met so many people in my life that it really is impossible to mention everyone who has shared many of my experiences. To attempt to write their names would be foolish – I would no doubt leave people out in error. So this is a big collective thank you to everyone who has been there with me over the years.

However, some people really do deserve special mention.

From my childhood, I would like to thank Mrs Bourne (Nancy) and the Bourne family. You were like my second family. Johnny Bourne, my childhood friend, died in September 1997. He was far too good a person to die so young.

Many birders supplied clarifications on dates and birding moments. So, my thanks go to Kieran Grace, Jim Fitzharris, Tony Lancaster, Brian Haslam, Steve Wing, Ron Johns, Jim Dowdall and Robert Vaughan.

I would also like to thank Anthony McGeehan, Paul and Andrea Kelly, Eanna Flynn, Pat and Bernie Kearney, Pat Corbett, John Donohoe and Davey Farrar for their friendship and support.

Derek Mooney, Eanna Ní Lamhna, Richard Collins, Terry Flanagan and Jim Wilson – thank you for the many great moments on the *Mooney Goes Wild* show. I would also like to thank Conns Cameras and the team at Swarovski Optik for their continued support.

John Fox, Gerald Franck and Philip Clancy (the lads), I promise that I will ignore the wing tags on harriers from now on and concentrate on plumage features for my identification criteria. Thanks for your friendship, lads.

Fergal Tobin, thank you for giving me my first opportunity to write for Gill & Macmillan. I would also like to take this opportunity to thank the whole team at Gill & Macmillan for their great work and support over the years. I would especially like to thank Deirdre Nolan for 'seeing' where this book was coming from. Thank you, Deirdre, for your encouragement and advice.

Several people very kindly read early drafts and offered constructive feedback that has greatly improved the contents of this book. So thank you, Anna Digby, June Digby, Sarah Carty, Anne Fox, Pat Corbett, Geraldine Boland, John Edmundson, John Fox, Gerald Franck and Irene Kavanagh. I would also like to thank Evelyn and Harry Johnston for their support.

Michael O'Clery has been a very special and true friend for many years. We have had many great birding adventures and experiences together over the years. He has been with me through thick and thin. Thank you for your friendship, Mick, and for the many great laughs we had along the way.

Della, Conor and all the O'Clery clan, thank you for making me feel a part of your family.

I would not be where I am today were it not for my own family. A big thank you to Paul and Clare – your encouragement, support and shared childhood memories are very much appreciated. Thanks also to Sylvia – you are still 'me favourit' cousint'.

My parents, Ann and Tom Dempsey, are the individuals who shaped me and made me who I am today (blame them – it's their fault). Words simply can't express my thanks to them for being such wonderful people and for being such wonderful parents. My mother assures me that Da would have loved this book. Thanks, Ma – that means a lot to me.

Hazel, you might appear last here but you are by no means least. Thank you for your patience and for holding the fort while I immersed myself in writing this book. (Even Suzie, our faithful spaniel, endured hours of sitting at my feet as I sat writing instead of playing ball or going for walks!) Without your love and support, this book simply would not have been written. Your honest feedback on early drafts was vital in seeing it through to completion. Thank you for your wonderful, amazing, extraordinary and fun company on life's journey.

Eric Dempsey
March 2015

PROLOGUE: A HIGH TIDE

What was this feeling of unhappiness?

The obvious answer was that I had just turned forty years of age and that this was a mid-life crisis. I have to admit that I don't buy into this mid-life crisis school of thought. We all have moments when we stop and take a good look at our lives. We all ask many questions of ourselves at different times, and I was, and still am, no different. Have I achieved what I wanted to achieve? Has my life taken the direction I hoped it would? Am I happy with my life? Are there things I would wish to change? What of the future? Is this the kind of life I want to continue to live?

These sorts of questions do not, in my opinion, reflect a crisis in life but rather a good long hard look at life, a 'stock take' for the want of a better phrase. For many people, the answers to these questions might all be positive. They might be living the exact life they've always wanted to live. For others, perhaps responsibilities of family or business mean that they have no choice but to plough on regardless of whether life has dealt them the exact hand they had wished for.

In August 2001, following an extraordinary four-week trip across the Tibetan Plateau, I returned to work and sat down in my office. I had a small team of people around me and it was nice to meet some friends and catch up. My boss was also very happy to have me back as we had lots of new projects for the autumn ahead. There were piles of emails awaiting my attention. There were meetings to go to. There were business plans to read over and prepare. There were more meetings. I sat at my desk and looked around me, and I wondered how in the name of God I had ended up sitting in the head office of Eircom.

However, my life was good. I had a well-paid, secure job. I had a house and a car. My job allowed me the luxury of travelling to these far-flung places on holidays. I didn't really want for anything. I was comfortable. Yet something just wasn't sitting right in my mind. Something wasn't sitting right in my heart. While others might satisfy this feeling with the purchase of a fast sports car or a motorbike, I had no desire for anything like that. I wanted to spend my life doing what I loved most – being out birding and enjoying nature.

I spent all of my spare time in the field watching birds, taking photographs of birds, sketching birds … I was talking, eating and sleeping my passion. The 'birdline', the rare bird information network, was keeping me right in the heart of the latest rare bird news. My name was well known by Irish and British birders: I was 'the man who ran the birdline.' I was engaging with other birders all over the country as I updated them with the latest sightings of rare birds each night.

I had also been very fortunate in that a book by Michael O'Clery and myself, *The Complete Guide to Ireland's Birds*, had been published in 1993. In many ways this book was my CV. RTÉ had even screened a short TV documentary on my birding life. So the birding side to my life was soaring. I was also busy working with Michael on a second edition of *The Complete Guide.* How lucky was I to be able to combine both childhood ambitions: I was birding and writing.

At weekends, I was busy with either guiding groups from the Tolka Branch of Birdwatch Ireland or, occasionally, guiding visiting birders who wanted to see Ireland's birdlife. I loved it. People were finding me through a website I had set up and this aspect of my birding was increasingly in demand. Bird guiding was big in the US and people were now travelling to Ireland to see birds.

The trouble was that I was doing all of this in my spare time. Work really was getting in the way of my life.

So these self-reflective thoughts about my career and my life dogged me. I felt that life was too short to simply stay within the accepted norms of a good nine-to-five job. I had tested the water and felt that perhaps guiding tourists might just offer me a way out.

Like everyone, I remember well where I was when I heard of the 9/11 attacks on the twin towers. I was at a meeting at work and news filtered through that something was 'unfolding' in New York. The meeting was abandoned and we all found a TV where we watched the events as they happened.

The shock of what happened on 11 September 2001 reverberated around the world. Everyone was speaking about it. TVs were showing (and re-showing) the events. However, what I had not predicted was that, within days of 9/11, several US birders who had planned to visit Ireland to go birding, with me as their guide, cancelled their trips. Suddenly and unexpectedly, it seemed that guiding was not going to offer that way out of work after all.

That autumn I threw myself into my birding. I had superb seawatches off Mayo where we witnessed huge numbers of migrating seabirds including at least 1,300 Leach's Petrels, 25 Sabine's Gulls, 1,000 Sooty Shearwaters and 10 Long-tailed Skuas all on one day. I saw a Stilt Sandpiper in Lough Beg, Co. Derry, a Hoopoe and an American Golden Plover in Co. Dublin and, best of all, a stunning Bluethroat in Co. Wexford. Autumn migration really is a great distraction from life.

It was mid-November 2001 when I got the first email; it asked if I would be available to guide a small birding party from the US around Ireland in the spring of 2002. The following day, I received another email, from a South African birder who was looking to hire me in February 2002. By the end of the week, I had received two more email enquiries. The following week, I had yet another, looking for my services in the summer of 2002. I was in demand again.

One Saturday afternoon in late November I went up home to visit Ma and Da (it's funny how I still considered 'home' to be the house where I was brought up). I decided to approach the subject with them. Even at forty years of age, I really valued their advice.

They were both astute and intelligent, and they always offered wise counsel.

I was in the kitchen having tea with Ma when I decided to raise the subject. I wasn't sure how she would react …

'Eh … just to let you know that I'm thinking of giving up the job and doing birding full time,' I announced.

Ma sat down and we spoke about all the 'ins and outs' of such a decision.

Was I sure I wanted to give up such a good job? This was her main question. She was concerned that I was rushing into making a decision that, in time, I might regret.

'There is no way back if you take that leap,' she said.

I understood her fears and really appreciated the fact that her concerns centred on my welfare and my future. She saw that I was deeply unhappy and not content with my life. Ma was (and still is) a very insightful woman.

'But if that's what you want to do, then I say go for it,' she said finally. 'But see what your father might think.'

Da was in the front room reading. I brought him in a cup of tea and sat down across from him. We chatted about all sorts of things until eventually I got to the subject I wanted to really discuss with him. Da was always very careful with his advice. He never offered advice until it had been well thought through. I felt slightly nervous of what he might say.

'Eh, Da … I would like your advice,' I started.

He listened silently with interest.

'I'm thinking of giving up the job to try my hand at birding for a living.'

He continued to listen.

'What'd you think … ?' I said.

Da remained quiet for a few moments. He was looking deep into my eyes. He was 'seeing me'. He was one of the few people who really 'saw me'.

'As Shakespeare once wrote,' he started. He smiled before continuing:

There is a tide in the affairs of men
Which, taken at the flood, leads on to fortune;
Omitted, all the voyage of their life
Is bound in shallows and in miseries.

Then he looked at me.
    'Is the tide high?' he asked.
    'Yes, Da … I think it is.' I replied.
    'Is your ship seaworthy, Eric?'
    'I think it is, Da,' I replied.
    'And have you planned your voyage?'
    I nodded.
    'Then set sail,' he said.
    He leaned towards me and put his hand on my shoulder.
    'But let me offer you some advice, Eric … As soon as you set
sail, always keep your eye to the horizon … never, ever look back!'
    I tendered my resignation the following day.

On 15 February 2002, I left my good career, walked out the door of
my office, took a big deep breath and set sail.
    Oh, by the way, Da, I'm keeping my eye to the horizon and I've
never, ever looked back.

Chapter 1 ⇥

HATCHING

I was born in the back bedroom of 23 Glasanaon Road, Finglas East, in north Dublin. It was 15 July 1961 and it was a warm Saturday morning. It had been a long forty-eight hours of 'hard labour' for my mother. With the help of a lady doctor (Dr Magee), I arrived screaming and kicking into this world just as the first rays of the rising sun crept above the roofs of the neighbouring houses. Ma informs me that the birds were singing just as I arrived but also quickly reminds me that she wasn't exactly singing for joy along with them.

It seems that I was 'a jaundiced baby' when I arrived, so much so that my cousin Sylvia took one look at me and commented that I 'looked like a dandelion'. Ma was very ill after she had me and, to this day, maintains she is still suffering from post-natal depression. From birth, having a thick skin and a good sense of humour was a prerequisite for survival in the Dempsey clan.

I am also reliably informed that I was a 'colicky' child and that I 'never kept anything down'. I also never stopped crying. In fact, I cried to such an extent that Ma apparently considered if there were any possibilities available to her to 'give me back'. She was a master in the art of bringing things back to the shops, but she'd lost the receipt for me.

I was given the name Eric Gerald Dempsey; Eric because Ma liked the name for some reason and Gerald after my father's brother. I was the youngest of three. Paul, my brother, is almost seven years older than me, and my sister, Clare, is two years older. Paul still questions whether I was born or if, in fact, I was hatched.

It would perhaps make for juicier reading if I said that my childhood was terrible before going on to regale you with the horror stories of my upbringing … but I can't. I have nothing but happy, wonderful and safe memories of my childhood. I know that I am lucky.

Our house was the second in a row of terraced 'corpo' (county council) houses. We had three bedrooms upstairs. The front bedroom was the biggest and it was where Ma and Da slept. The back room was small and it was where Paul and I slept when I was old enough to move in with Paul. Clare slept in the box room, which barely had enough room for her bed and a set of drawers.

It's funny, but, as a child, I thought our house was huge. Walking back into the house now, I realise how small it really is. Even the 'big' front bedroom is very small.

There was a tiny, square landing at the top of fourteen narrow stairs that led up from the hall below. Above the landing was the entrance to the attic that required Olympic gymnastics skills to climb into (we didn't have a ladder for a long time).

Downstairs, when you came in the front door, there was a narrow hall with 'the front room' on the right. Again, this was the lap of luxury in my childhood but it is a very small, cosy living room. We had the telly in the corner near the plugs. We had a tall light stand there too. There was a large table that was given to us by a relative (I don't know who), with four chairs. We ate here because the kitchen was too small. We also had a big old-fashioned wireless (radio) in the corner that had a big knob which was turned to tune in to the world. Ma would listen to *The Kennedys of Castleross* each day on Radio Éireann. As a child, I used to think there were tiny people living inside the wireless. Being just one or two years of age, I obviously didn't grasp the concept of radio waves (I'm still not very technically minded).

Passing the front room, you entered the back of the house – it took about three steps to get from the front door to the back of the house. Here, there was a narrow hall where we had a tiny toilet on the right and then a separate bathroom on the left.

We also had a small 'press' on the left where Brandy, our dog, had her bed. It was her little corner of the house. The press was also where the gas meter was kept. This meter took old shilling bits (old Irish shillings had an image of a bull on one side and a harp on the other). The gasman would arrive every other Saturday morning and he would go to the meter. There, he'd check the reading, unlock the meter and empty out all the shillings onto the table. I was always fascinated by the speed at which he counted all of the shillings, sliding them quickly into his hands and placing them in mini-columns on the kitchen table. Ma had a foreign coin (I think it was an old French franc) that she used in emergencies if the gas was running out and she didn't have a shilling to hand. I was always impressed at how this gasman would, with the slightest flick of one finger, eject that coin from all the shillings. It was a skill of many years counting money. Having checked the reading, he would know how much he should have and would then leave a few shilling coins on the table (along with the franc) for Ma to use again.

Just past the press, on the right, was a tiny L-shaped kitchen. It had a big old enamel sink under a narrow window. The house was slightly renovated when I was very young so that we had a combined toilet and bathroom as well as a slightly bigger kitchen. From then on, all of our meals were eaten in the kitchen around a small table. The big table in the front room was passed on to a relative who needed it and we got a sofa.

More important for us kids was the green back door that led into our back garden. Again, as a child, this garden, only 13 m in length and 3 m wide, seemed as vast as the Serengeti. This was where I was sure I heard lions roaring at night in the honeysuckle that grew along the top of the end wall. This was where I scored a winning goal in the FA Cup Final at Wembley. This was where I camped in the remote jungles of the Amazon. This was where I shared my first dawn chorus with Da. Our garden was our playground both in the physical sense and for our imagination.

Our back garden also had a coal shed. I remember the coalmen arriving on our road. I know that, for some, the following recollection might make it seem like I was raised in the Victorian age, but the coalmen arrived along the road on open horse-drawn carts where sacks of coal were lying one against the other. You'd hear them a mile away.

'Cooooaaaalllllll!' they'd shout, with an upward inflection to the end of the word.

It was shouted in a mighty Dublin accent. Now that I think about it, I remember that these men shouted in the same accent and with the same volume as the men who sold newspapers. As a child, I never understood what these newspaper men were shouting when they'd roar:

'Herri dah Press … Press ah Herri!'

It was only later that I realised they were selling copies of either the *Evening Herald* or the *Evening Press* ('*Herald* or *Press* … *Press* or *Herald*!').

When the coalmen came to deliver coal, I would stand back and admire their strength. Panting heavily and hunched over, they lugged heavy sacks of rough coal through the hall before emptying the sack into the coal shed. I struggled to even lift a full bucket of coal. These men were covered in soot and dirt, but they were like ancient warriors in the strength they seemed to possess.

The coal shed was also where I kept my pet mice when I had them and, later on, Sheba, our guinea pig.

We had a small front garden: we had a small lawn and flowerbeds and, walking from the front door down the narrow path, you took one step down to the gate and railings. The gate led onto a busy road where the number 19 and 34 buses raced up and down. There weren't too many cars around in the 1960s, so, once you kept your eye open for buses, the road was another playground. There were also small greens (small green areas) where we could play. My best friend, Johnny Bourne, lived across the road from me, right in front of one of those greens. The Bournes were like my second family and Mrs Bourne was like another Ma. If Johnny Bourne wasn't with me playing in my house or garden, I was over there playing in his.

We were very lucky in that there were shops right beside us. We had everything, including a greengrocer, sweet shops and newsagents, a butcher and a chemist. We also had a pub called the Red Dragon right across from us (its name was later changed to the Cremore House). Around the corner was another pub called the Quarry House. Finally, there was also a chipper where we sometimes treated ourselves, but that was a rare treat.

'Don't let her give you reheats,' Ma always warned the person tasked with going to the chipper. 'Tell her you want fresh chips.'

Every morning, I would hear the clink of full milk bottles being left on the doorstop and the rattle of empty bottles as they were collected. This sound was like the Finglas dawn chorus. In the days before 'recycling' became the buzzword, the constant reuse of milk bottles was a superb example of what recycling should be. It was the same for jam jars – a man used to call to collect those – and lemonade bottles, where the shops would give three old pence for each bottle returned.

The milk was delivered in the early hours of the morning so that we all had fresh milk for breakfast. Our milkman was Eddie. He was a grey-haired man with a small moustache. He arrived on a horse-drawn float. This was also open but had a little cabin where he sat. The horse would slowly amble his or her way down the road as Eddie lifted full crates of milk, placed the milk bottles on the doorsteps and filled the crate with empty bottles. The crates would be stacked up high on the float.

We used to leave little egg cups out on the empty milk bottles. Eddie would place these on the tops of the full milk bottles. It stopped the Blue Tits from attacking the foil caps, peeling them back, and drinking the cream from the top of the milk.

Ma didn't like Eddie that much because he wasn't nice to his horse. If the horse hadn't caught up with him, or went too far, Eddie would give the poor old horse a slap in the face. If Ma caught him 'mistreating' his horse, she'd give him a good ear bashing. Eddie didn't know that Ma was (and still is) a long-time campaigner on animal welfare issues. Animal cruelty was not tolerated in our home.

So, from delicate threads found around and within the bricks and mortar of my birthplace, the tapestry of my character was woven. Home was where my thoughts, integrity and principles were shaped as if from wet clay by the skilled potter's hands of my parents. Home was where I learned to laugh. Home was where my imagination was allowed to run riot. Home was where I was allowed to dream. Home was where I emerged from to be the man I am today.

I often try to recollect my first memories. It's a hard thing to do, to remember when you first became conscious of your surroundings to the point where they formed an actual memory you stored in your brain.

I do remember being in a cot beside my parents' bed. I remember standing up in my cot and Ma lifting me out of it and bringing me downstairs. I have no idea why I have this particular memory. Perhaps it was the journey down the stairs that made the experience stick? I also remember sitting in an old-fashioned pushchair. We were leaving Clare to school and, on the way home, Ma asked if I wanted to get out and walk. I walked along for a while, holding on to the side of the pushchair. I remember that day because I saw a dead dog that had been killed by a car. I cried all the way home.

However, amongst the myriad of very early childhood memories is one that really does stand out. I can see it so vividly that I sometimes doubt that it really happened. Thankfully, Ma confirms that my memory is accurate. I was probably no more than three years of age (possibly younger). I was at home with Ma. We were in the kitchen and Ma was preparing some food for me. She was also concentrating on looking out the window into the back garden.

'Here … come look at the birds,' she said, as she took me into her arms.

She lifted me up and I stood on the big enamel kitchen sink so that I could see out of the window.

'See the mammy bird feeding her baby,' she whispered to me. 'Look at the little baby bird flapping his wings.'

I looked out the window and there, in the garden below, was a House Sparrow with two chicks. The chicks were begging and doing their usual wing flapping and 'feed me now' open-mouthed posturing. I was transfixed.

I am sure that I must have seen birds before. I know Ma and Da were always keen to show us the birds in the garden or to get us to listen to their songs. However, for some reason this was the first time I can definitely say I became 'aware' of birds.

Ma put me down and went on preparing food. I begged to be lifted back up so that I could continue to watch the birds. According to Ma, we stayed looking out that window for about an hour until it was time to go out to collect Clare from school.

Apparently, I could not take my eyes off this little family of House Sparrows. I suppose I can say without a shadow of a doubt that House Sparrow was technically the first bird species on my life list.

For many years after, Ma would jokingly say that she regretted ever lifting me up to look at those birds. Come to think of it, she still says that to me on a regular basis.

'Look at what I started!' is her usual quip as she rolls her eyes to heaven.

It is always said with a smile. Paul, on the other hand, maintains that this fascination I have with birds is just a phase that I will eventually grow out of.

However, from that first bird-awareness experience on, I was fascinated by nature of all sorts. I would look at butterflies and ladybirds; I would watch birds and listen to their songs. I remember

Da bringing home a Hedgehog. I couldn't believe my eyes. Da said he found it in the middle of the road close to our home and that it would have been killed if he hadn't stopped his bike and picked it up. He put it in the saddlebag of his bike and brought it home to show us all. We fed it some dog food that Ma rushed down to the shops to buy. The Hedgehog was not put out by the experience (nor indeed by being in our company), and wolfed into the food. We then put him in a cardboard box and we went for a walk to the local fields where we released him. It was another memorable early wildlife experience.

I started school in September 1966, when I was five years old. I went to the Sacred Heart Boys' National School on Ballygall Road East (just around the corner from our home). It was a prefab school.

I couldn't wait to start school. I had a little brown schoolbag with my copybook, pencils and schoolbooks. I remember, on my second day of school, spilling my little bottle of milk inside the bag and the milk going all over my books. To this day, the stale smell of spilt milk reminds me of that little schoolbag.

Ma brought me around to school on that first day and a rotund 'baldy' man with glasses met us. He was Bunny McCarthy, the headmaster. He frightened me a little.

I entered the world of education under the tutelage of Mrs McCarthy (Bunny's wife). She was a plump, tall countrywoman with grey hair swept back into a net, a round and gentle face and a welcoming smile. She wore glasses that, more often than not, sat on the end of her nose during the day. I was in 'low babies' (to be more politically correct, Junior Infants). She taught me for two years, in both low and high babies, and they were the most brilliant two years of my life.

From the very first moment, I found a kindred spirit in Mrs McCarthy: she loved nature too. She told us all about birds, flowers, butterflies and hedgehogs. She brought us out to look at birds and showed us trees and flowers, butterflies and spiders. She even taught us how to tell the difference between the songs of a Blackbird and a Song Thrush ('The thrush always sings his song twice!'). She had a nature table. Each day, she would spend time speaking to us all about nature and would bring something new into the class for us to look at. One day it might be a shell, the next day it could be a feather. In autumn, we went collecting leaves and we'd try to tell which leaf was from an oak tree and which was from a beech. It was an education that changed my life.

A nature table is such a common thing to see in school classrooms these days, but in 1966 it was a real rarity. Nature studies were not part of the school curriculum. I was so lucky to have been assigned to this woman's class. I often wonder what path in life I might have followed if I had been put into another teacher's class on that first day of school. In fact, I was doubly lucky: not only did I have parents who were interested in nature and encouraged my interest in the natural world, but I also had Mrs McCarthy as my teacher.

As I reflect on the great influence this one teacher had on me, I wonder just how many others she might have influenced. Amazing as this might sound, I do know that up to seven of Ireland's top birders all have one thing in common – like me, they also had Mrs McCarthy as their teacher in low and high babies. I am reliably informed that several of Ireland's leading botanists also had her as a teacher. To say this woman influenced whole generations of young fellas from Finglas, Glasnevin and Ballymun would not be an overstatement.

Now, when I stand in front of a classroom of kids in a national school, sharing my love of birds and nature with the birdwatchers of the future, the romantic soul in me allows me to believe that I am somehow carrying the baton for Mrs McCarthy. She is a hard act to follow.

Chapter 2 ➵

ON BEING A SINNER

'I was born with the stain of original sin on me soul.'

I was in high babies when Mrs McCarthy told us about original sin for the first time. We all loved Mrs McCarthy but today she was frightening the life out of us. She looked down at the class over the rim of her glasses and told us about Eve eating God's apple and how because of this we all had a big stain on our souls called 'original sin'.

And it wasn't even our fault.

'No, it was all Eve's fault 'cause she ate an apple.'

I wondered if it was a Granny Smith … I loved Granny Smiths. I loved how crunchy and juicy and sour they were all at the same time. Then Mrs McCarthy showed us a picture of Eve eating her apple. It wasn't a Granny Smith, because they were green. The apple she was eating was a big red one.

'And there was a snake beside her watching her eating the apple. We didn't have snakes in Ireland 'cause St Patrick got rid of them for us. And the snake was the Devil. And the Devil was making her eat the apple … and that was a sin!'

My six-year-old mind was running riot.

'But then she made Adam eat the apple and he committed a sin.' (I didn't know what 'committed' meant but it had to do with sins so I decided I'd never commit a thing in case it might be a sin.) 'You see, this red apple was God's and Eve stole his apple … so this was a sin and because of that I was born with a big stain of sin on me soul.'

I was sure that I could feel the weight of this sin on my soul. I was terrified.

'And because they had eaten the apple they were banished from the Garden of Eden.' (I had no idea what 'banished' meant but it couldn't have been good either.) 'This garden was lovely and had loads of trees and animals everywhere. There were also loads of birds flying around it.'

I thought that the Garden of Eden might be a bit like the Botanic Gardens, or 'the Bots', where Da took us every weekend … but maybe it was a bit bigger and with lots more birds to look at.

'Because Adam and Eve were banished from the garden, they didn't have any birds to listen to or look at, or flowers to smell. Imagine what it would be like to have no birds to look at!'

And even worse: 'Because Eve had stolen an apple, I mightn't get to Heaven if the angels come in the middle of the night to take me to God!'

It was all too much for me. However, just as panic was beginning to set in, Mrs McCarthy restored calm.

'But', she said, 'because all of your mammies and daddies had you baptised, that stain of original sin has been wiped from your souls'.

The relief her words brought was overwhelming.

'Because Ma and Da had me baptised, I won't end up in Purgatory, or Limbo, or some other place where children go who were born with the stain of original sin but whose mammies and daddies didn't have their souls cleaned by getting them baptised.'

Ma and Da have saved me soul, I thought to myself.

I knew for a fact that I had been baptised, because I had seen pictures of my Aunty Eva outside the Tin Church holding me as a baby. I couldn't see 'me' in the picture but Ma told me that it was definitely me. Aunty Eva was my godmother and because I had a godmother that proved I was baptised.

Aunty Eva died when I was just a baby, and Ma told me that from that moment on she was my guardian angel and that she 'watched over me every day'. I didn't understand what Ma meant by 'watching over me'. As far as I knew, Aunty Eva just 'watched' me every day. I reckoned she sat on my right shoulder, though I never, ever saw her. And because she watched me every day I had to be careful because she was Ma's sister and if I did anything wrong she always told Ma. That's how Ma always knew if I'd been up to something – Aunty Eva was a bit of a tittle-tattle.

Just as I was beginning to feel happy that the stain of original sin had been wiped off my soul, Mrs McCarthy dealt another fatal blow.

'But,' she announced, 'since you were all baptised, you've all committed loads of other sins.'

Apparently, in the six years since I'd been baptised, my soul had become black with the sins I had committed. I was sure Aunty Eva was keeping track of them all. But how could I get rid of these sins?

Mrs McCarthy provided the answer: we were all going to make our First Confession. We were going to confess our sins (I didn't know what 'confess' meant but I knew that it had to do with sins, so it couldn't be a good thing). We were making our First Confession because we were getting ready to make our First Holy Communion.

I understood it all very clearly. We would get to taste the body of Jesus in a bit of bread the priest would give us when we made our First Holy Communion. But we couldn't taste the body of Jesus if we had loads of sins on our souls. So we had to go to confession to get rid of those sins first.

The trouble was that, no matter how hard I tried, I couldn't really remember sinning. But if Mrs McCarthy told me that my soul was black with sins then it had to be. Mrs McCarthy knew everything.

Weeks of rehearsals passed. We pretended to be telling our sins to the priest. We practised our prayers and I knew that when the day came I'd be ready to confess all of my sins. But I still couldn't think of any sins I'd committed and I was starting to panic that I wouldn't have anything to tell the priest about. The day before my First Confession, I decided I'd better sin a bit. So I told a lie to Ma.

'Have you washed your hands?' she asked before I sat down for dinner.

She always asked Paul, Clare and me that question.

'Yes,' I replied … but I hadn't washed my hands.

At least now I had a sin to confess.

After dinner I went out to play with Johnny Bourne. He was a year older than me, so he was in First Class, but he was also making his First Confession and First Holy Communion like me. We kicked football for a bit and then decided that, as we had a few pence between us, we should go to the shops. By next week we'd have loads of money because we were making our Communion and people always gave you money when you visited them.

It was a nice afternoon as we strolled back with a bag of Perri crisps between us. As we passed McQuaid's shop, we stopped to look at all the fruit and vegetables in boxes outside. There was a big crate of red apples.

'Eve would love those,' I thought to myself.

The colours in the display were lovely: the green cabbages and the orange carrots; the red apples and the yellow bananas; the red tomatoes and the brown potatoes. I liked the colours, especially when they were all lined up beside each other.

'Raw peas are lovely to eat,' Johnny announced.

His Ma and Da grew vegetables in the garden, so he knew these things. His words filtered into my brain: 'Raw peas are lovely to eat.' I had eaten tinned peas; Bachelor's peas were the best. But raw peas?

'You mean those yokes?' I asked, pointing to a load of garden peas in their pods sitting in a small basket on a stand.

'Yeh. They're like sweets. You open the package and you can eat them.'

I looked around. There was no one in sight. We had spent all our money on crisps, so we couldn't buy peas.

'I've never eaten them,' I stated, 'but I'd love to try them.'

'You can always take one,' Johnny urged. 'I'll keep watch.'

Within a second, I had a pod in my pocket and we both walked quickly to Johnny's back garden, our hearts thumping in our chests. There we opened the pod and found four little peas inside. We ate them. They were sweet and tasted really, really nice.

The following morning, all of the high-baby classes and some of the first classes were lined up in the schoolyard and, in orderly lines, we were marched down the hill to the Tin Church. This was our big day. This was the day when all of the black sins we had committed since we were baptised would be wiped off our souls. As I walked along, I practised the lines over and over again.

'Bless me, Father, for I have sinned. This is me first confession.'

After that, I'd have to tell the priest my sins. Then I'd say, 'For these and all me other sins I am heartily sorry' (I didn't know what 'heartily' meant but it was to do with sins so I swore I'd never heartily anything in case I sinned).

Then I had to say an Act of Contrition (I didn't know what that meant either but if it helped to clean my soul of sins I would not forget to say it). After that, the priest would give me penance before he'd let me out of the confession box. Penance was a load of prayers he told you to say to prove to God that you were really sorry for all your sins. I had it all off by heart: I was ready for confession.

We sat in lines near the confession boxes. There were four boxes, one in each corner of the church. Each box had two little doors, one at each side. In the middle was another big door with a red curtain in front of it. The priest sat on a seat behind that curtain.

I sat looking around me, waiting my turn. The Tin Church was always cold, even on hot summer days. The interior was pale and grey and the wooden seats were hard. I looked at the statues. The Blessed Virgin was standing looking down at me. And on the wall was a picture of Jesus on the cross. Mrs McCarthy told us that he had died to save us from our sins. This sinning stuff was very

serious. I mean, Jesus died because we had all sinned! At the top
of the church, near the altar, candles burned close to where a few
aul' wans prayed with rosary beads in their hands. At each of the
confession boxes sat us 'first confessioners'. Teachers guided each
young fella into a box and then, when he was finished, guided him
to where he should pray to God to prove he really was sorry for his
sins.

I watched each of my classmates go into the box, but I watched
them even more closely as they came out. They didn't look any
different, but of course I couldn't see their souls, which were now
pure white because they had their sins wiped off them.

Slowly, as each of my classmates went into and out of the
confession box, I edged closer and closer to the aisle. I practised
my lines over and over: 'Bless me, Father, for I have sinned. This is
me first confession.' From where I sat, I could hear the mumbled
confessions of the others. I looked up at the picture of Jesus on the
cross dying for our sins.

Then, Mrs McCarthy tapped me on the shoulder. It was my turn.
I went into the right side of the box and entered into the darkness.
I knelt down, looking up at the small wooden hatch in front of
me. It was behind a wire cage. I supposed the priest needed to be
in a wire cage to protect him from all those sins that were flying
around the place. The inside of the box had a strange smell. It was
a mixture of polished wood and leather from the small little pad
that I was kneeling on. Even in the dim light I could see that the
pad was worn thin from all the knees of countless sinners like me.
Was this how sins smelt? From within the walls of the confession
box, I could hear someone talking. It sounded like Mick Duffy.
I didn't like Mick Duffy. He pushed me over in the schoolyard a
few weeks before and I wondered if he was telling the priest about
that. I was sure that was a big sin and he'd better remember to tell
it because the sin would stay on his soul if he didn't. The talking
stopped.

Suddenly the wooden hatch slid quickly back and the dark
outline of a priest appeared. I could just about see his face but I

couldn't be sure which priest he was. He didn't need to say anything: that door sliding back was enough to spring me into action.

'Bless me, Father, for I have sinned. This is me first confession.'

It was a confident start and I was happy that I had remembered my lines.

'And what sins have you committed?' the priest asked.

'I told a lie to me Ma,' I replied. This was the sin I had lined up yesterday.

'Lies are very bad. You should never tell lies.'

'Yes, Father,' I replied quietly.

'And have you any more sins to tell me about?' he asked.

He sounded nice. His voice was friendly. I didn't want to let him down so I told him I'd been bold a few times. I didn't really remember being bold but felt that maybe this would be enough of a sin to keep him happy. Then, just as I was about to say, 'For these and all me other sins, I am heartily sorry,' I remembered the peas.

'Oh', I added. 'Eh, I stole too!'

Silence consumed the inside of the confession box. Total silence. It seemed to last an hour. Maybe he didn't hear me? Maybe he'd gone asleep? Hearing all these sins must have been very tiring. I was about to say it again when the silence was broken by a long, deep sigh.

'You stole?'

'Yes, Father.'

'What did you steal?'

'Peas, Father.'

'Peas?'

'Yes, Father.'

'You mean you stole a tin of peas?'

His voice now seemed angry. This didn't feel good. I sensed I was in deep trouble. Why did I mention the peas?

'No, Father, I stole ordinary peas.'

'Ordinary peas? What do you mean by ordinary peas?'

By now I was breaking into a sweat. He was getting very angry with me. His voice was getting louder. I was sure Mrs McCarthy was listening outside and I'd be in big trouble when I was done here.

'I mean they were raw peas, Father,' I tried to explain.

'Raw peas? What do you mean by raw peas?'

'You know,' I answered. 'They were peas in a green yoke.'

'A green yoke … Do you mean in a pod?'

'I don't know, Father'.

This was the truth – I didn't know what a pod was.

'There is a green yoke,' I explained. 'When you open it, there are raw peas in it.'

'That's what I said,' he stated slowly and deliberately, making sure I understood each word as he spoke them. 'You stole a pod of peas.'

'Yes, Father,' I conceded.

If he told me that the green yoke was called a pod, then I wasn't going to argue with him. At this stage I was in a panic. No one else had been this long in confession. Surely Mrs McCarthy would be wondering what was keeping me. She must know by now that I had a serious big black stain on my soul. I was in big, big trouble. And Jesus had died to save me from sin and here I was with this big black sin on my soul and the priest had discovered it.

'Do you know that when you steal you're doing the Devil's work?' the priest said.

'And,' he added, 'when you stole those peas, the Devil was whispering into your ear!'

I was now terrified. I didn't remember the Devil whispering into my ear but I did remember Johnny Bourne encouraging me to take the peas. So if the priest said it was the Devil telling me to do this then Johnny Bourne must be the Devil … and I thought he was my best friend.

I had visions of the snake telling Eve to steal that red apple. I must have been committing an original sin just like her. And she'd caused a lot of problems because we all had to be baptised because of her. What had I done? This was really serious.

'And when you listen to the Devil, your guardian angel cries and feels pain,' the priest then announced.

Now I was in even bigger trouble.

My mind was reeling. I had made Aunty Eva cry and feel pain and she was going to tell Ma about it and Ma'd be really annoyed with me. I found I was shaking, my arms, my legs, my whole body. There was no escape. The Devil had a hold of me and I was going to Purgatory or Limbo or some other place with all the other children who had the stain of original sin on their souls.

As these thoughts were racing through my mind, I realised that silence had once again descended. I took the opportunity to escape.

'For these and all me other sins, I am heartily sorry,' I said loudly.

The priest muttered something in a language I didn't understand, or maybe he was still giving out about me under his breath. I didn't care. I launched into an Act of Contrition. When I finished, the priest leaned close to the wire cage until his nose and face were pressed against it. I could see he had large hairs growing at the tip of his nose.

'Say five Hail Marys and a special prayer to your guardian angel,' he said. 'And never listen to the Devil again!'

With that, he made the sign of the cross at my face and the door slid shut. I was plunged into darkness again.

I stumbled out of the confession box sure in the knowledge that the whole church would be looking at the greatest sinner ever to have made a confession. But no one was looking. Mrs McCarthy guided me to the next row of seats where I knelt and said five Hail Marys. I also said a prayer to Aunty Eva.

'Angel of God, my guardian dear,
To whom His love commits me here:
Ever this day be at my side
To light and guard, to rule and guide.
Amen.'

In fact, I said it twice, hoping that she wouldn't tell Ma about me stealing the peas. I also hoped that she wouldn't tell Ma that Johnny Bourne, my best friend, was the Devil.

Later that afternoon, I passed the fruit and vegetable shop. The red apples were still there. So were the peas. I now knew how easy it was for Eve to have taken God's red apple, with the Devil whispering in her ear. I looked at the peas in the small basket outside the shop. The Devil told me to steal one of those peas and because of that I had almost brought the stain of original sin back into the world.

'But,' I reassured myself, 'that sin is gone now 'cause I've made me First Confession.'

Chapter 3 ⟿

THE GREATEST OF GIFTS

Christmas and the build-up to Christmas was always great fun in our house. Every year until we were teenagers, we had the excitement of getting the tree down from the attic on 17 December, as that was Ma's birthday. We had a false tree because real trees 'left a mess' (according to Ma). We'd pull the false branches out to make it look real. Then we'd decorate it with old Christmas balls, old Santys and tinsel. We'd put a silver star on top and Clare and I would finish with a cardboard Santy (mine) and a fairy (Clare's). Both had a paper pull-out section so that the fairy had a puffed-out dress, while Santy had a pull-out sack of toys. The tree stood in the corner near the telly and the plugs. When it was finally decorated, the lights would be plugged in and we'd all hold our breath to see if they worked. Even though the lights were ancient, they always worked.

Once the tree was up, Christmas had arrived. We all especially loved Christmas Eve. It was such an exciting day. Letters would be written to Santy and the thought that he was coming that night was almost too much for me. The atmosphere was added to by Ma's love of Charles Dickens and *A Christmas Carol*. For the week coming up to Christmas, we'd read sections of the book aloud after dinner each evening.

Santa Claus is of course the greatest and most wonderful scam played on children. The amazing thing is that, when you eventually do find out the truth, it's as if you understand the magic of believing and so become part of perpetuating the whole scam. I remember my brother, Paul, telling me that he saw Santy flying with his reindeers in the sky. I fully believed him. I was no more than five and he was twelve, yet he was as convincing as Ma and Da were.

Christmas was a time for buying bottles of cream soda and red lemonade. It was a time for going over to Mr Heaslip, who had a small off-licence, to help Da bring home some bottles of beer and a small bottle of whiskey for visitors who might call. It was a time of listening to Da play 'Silent Night' on his mouth organ, accompanied by Clare on her melodica. It was a magical time.

Going to bed on Christmas Eve is perhaps one of the most exciting experiences of a young life. Santy left the presents at the end of our beds, not downstairs at the tree. The sheer excitement of waking up in the middle of the night and seeing a collection of presents at the end of the bed is indescribable. There might have been a jigsaw and a selection box, as well as a tube of Smarties or Fruit Pastilles. And, if we'd been very good, Santy would have brought each of us the present we'd hoped for. Then we'd wake up poor Ma and Da so that they could see what Santy had brought.

I have so many memories of waking up on Christmas morning, I could write a thousand pages. However, one present does stand out. I remember that Christmas morning as if it was yesterday. I woke up to find a rocking horse. He (I decided the horse was a 'he') was magnificent: white with black spots, he had red reins and a saddle, a long black mane and tail, and red wooden rockers. I could not believe my eyes. He was just like a Red Indian's horse. It was beyond my wildest dreams because I was always an Indian and not a cowboy. I got a wigwam for my birthday the following summer.

It was years later that I found out how much poor old Da suffered in getting this rocking horse for me. Da worked as a porter in the Mater Hospital and, as the horse was so big and difficult to hide, he stored it at work. On Christmas Eve, when we went to bed, he got the bus down to the Mater and collected the rocking horse. It weighed a tonne. He lugged it from the Mater up to the bus stop at Phibsborough with the plan of catching the last bus home. It was here that his plans were scuppered. As he approached the bus stop, he saw to his horror that the whole of the Redmond clan were at the stop. These were our next-door neighbours and they had lots of kids who were going home in anticipation of Santy visiting. They had been visiting their grandmother (Nana Kirwan) in Cabra and were waiting to get the last bus home too. If he arrived at the bus stop with a big rocking horse then he would ruin it for the kids: they would instantly know that Santy didn't exist. He couldn't approach the bus stop. He watched in despair as the last bus home came and went. He had to walk the two and a half miles home, lugging a big rocking horse all the way. He told me that the joy on my face when I awoke on Christmas morning was worth all the effort.

I named this rocking horse Charlie and for many of my early childhood years I rode that horse across many western plains pursued by shooting cowboys. I galloped around Finglas on Charlie and was always accompanied by Brandy, my faithful dog. She was my Rin Tin Tin. Brandy was just a puppy when our other next-door neighbours, the Reids, got her for the kids at Christmas. Within a few weeks, this little blonde ball of fluff was at our door as the novelty of a puppy quickly wore off. It was love at first sight for Ma, and our neighbours were only too happy for her to move in with us. She was my best pal. She was part of our family. She was one of us. She played football and cowboys and Indians with us, went on walks with us, ate ice cream with us (she always got a cone when the ice-cream van came, but never got the ripple put on it). She was a patient and loving dog. She was a Dempsey. When she was put down in 1977, it was the first time I ever saw

Da break down and cry. Ma cried for months. In fact, I believe that Ma never got over losing Brandy. We, as a family, never got another dog.

I mention all of these memories because I recently discovered a box full of old letters, newspaper clippings and old photographs that Da had stored away in the attic. Among one collection of letters, I found a white envelope. Upon opening it, I came across two unexpected little treasures.

Inside were two little yellow handmade envelopes. Da always brought home pieces of yellow paper that were thrown out in the Mater. We would spend hours drawing or writing on these. I couldn't believe what I was seeing. The envelopes were both addressed to Santy in my childhood handwriting.

The first was addressed: 'Santy, from Brandy'. It had a small 3d stamp drawn on it (3d being three old pence). I opened it. It read:

Dear Santy,
  Please bring me a box of dog chocolates, a new ball, a new basket for a bed and a Christmas stocking.
  Love, Brandy
  Xxx

The second letter was addressed: 'Santy Claus, from Eric'. It also had a 3d stamp drawn on the envelope. I opened the letter with real excitement. It read:

Dear Santy,   Christmas 1968
  Please bring me a tractor and if you cannot, please bring me a track for my cars, a bell, lego and a lot of books.
  Eric Dempsey xxxxxx

I sat reading this letter from my seven-year-old self. I remembered wanting a tractor for some reason but Ma telling me that Santy mightn't have enough room in his sack for it so I should think of something smaller. Santy (in his wisdom) brought me a wonderful track for my cars. I have no idea why I wanted a bell, but I got a nice brass bell that sounded lovely. Lego was always great. However, it was the last line of the letter that really struck a chord. Seven-year-old me wanted 'a lot of books'.

I grew up surrounded by books. Ma and Da were avid readers as were Paul and Clare. I remember it was always said in our home that 'you'll never be lonely when you have a good book to keep you company.'

It's strange but I remember the moment I first realised that I could read. In fact I remember my 'breakthrough' word so clearly. I had a kid's book and I was slowly making my way through a sentence when I came across the word 'always'. It was a big word. I have no idea how young I was. I looked at the word and said, 'Always'. Ma, who was helping me with my reading, congratulated me on being able to read such a big word. It was the moment the world suddenly opened up to me. From then on, I wanted to read.

I read all sorts. I loved the comic annuals that came out at Christmas, I read storybooks and I read books about nature. However, early in 1969 my direction in life was sealed for ever.

Each week I was treated to a comic. My comic of favour was *Whizzer and Chips*. I was no *Beano* man. I would go to Dagge's, the local newsagent, to collect my comic. Mr Dagge was a tall, refined man who, in later years, referred to me as 'Mr D'.

'Good morning, Mr D, and how are you today?' was his usual greeting.

'I'm very well indeed, and how are you, Mr D?' was my usual response.

Leonie was the lovely shop assistant in Dagge's and she always kept a copy of *Whizzer and Chips* for me. This one particular week was no different. I went into the shop with Ma and, as I collected my comic, a new publication on the shelf caught my eye. It was an issue of *Purnell's Illustrated Encyclopaedia of Animal Life*, which was one of those weekly magazines that you collected over a few years to form a complete series of illustrated encyclopaedias. I opened it up and looked through the pages. There were pictures of birds, mammals, insects, fish and reptiles, all in alphabetical order. It had loads of detailed information about each animal.

I was hooked from the very first minute I picked it up. Here at last was a magazine that would tell me everything that I ever wanted to know about wildlife in the world. I picked it up and showed Ma the issue that was on sale that week. Apparently it had started a few weeks before that. I looked at Ma with pleading eyes.

'Can I get this too?' I asked hopefully.

Ma shook her head. 'No, you can't have both *Whizzer and Chips* and that,' she said. 'Which one do you want?'

For the first time in my childhood years, I abandoned my *Whizzer and Chips*. Ma thought that the 'animal' magazine was a little too advanced and that perhaps I would return to *Whizzer and Chips* the following week. However, when I got that magazine home, I read it slowly from cover to cover. It showed animals I had never heard of. There were birds with amazing colours and names. The spark that was already there from a young age was ignited that week. I wanted to read more. I wanted to learn more. I wanted to travel the world and see all of these animals and birds. I could hardly wait until the next issue. *Purnell's Illustrated Encyclopaedia of Animal Life* was the catalyst that gave me a sense of wonder about the natural world that has never left me.

The following week, Ma duly changed my order from the *Whizzer and Chips* to *Purnell's Animal Life*. Leonie always had a copy of it set aside with my name on it. Ma even ordered the back copies that I was missing so that I would have the complete set. As a real treat, when I'd collected each volume, she would send off for the binder covers that kept each magazine in its place. I collected all six volumes and read them from cover to cover. As I sit and write this, the complete set still has pride of place on the shelf in my study. These provided me with my first real insight into the wide world of nature as a young seven- or eight-year-old.

By the time I reached Volume 2 of *Purnell's Animal Life*, I was a nature and wildlife addict. Evidence to support this comes in the form of another little note written on that yellow paper that I found among Da's stored bits and pieces. It was from a nine-year-old me and was a simple reminder for him to set his alarm clock so that we could get up to hear the dawn chorus. Da woke me up in darkness and together we sat drinking tea outside and listening to the slow awakening of our garden birds. It was my first dawn chorus. I will never forget it.

We sat out on deck chairs with blankets around us, since it was chilly. Then, just as we saw the first glimmer of light in the eastern sky, a Blackbird started to sing from the very top of a TV aerial near the shops. His song was so clear. There was not another sound. Then another Blackbird answered him from across the road (I reckoned it was singing from Johnny Bourne's roof). It was the start of the gradual awakening of Finglas birds. A Robin sang from the trees in the garden over the wall at the back of our house. A Song Thrush soon joined in and Woodpigeons quickly added their 'coo-co-co-coo-coos' to the chorus. Slowly the sound built up with Rooks and Jackdaws, Starlings and House Sparrows until our garden was a cacophony of bird sound. The last bird to start singing was a Dunnock. There were Dunnocks nesting in our small hedge.

It was just Da, me and the birds. It was magical. I imagined that we were in the Amazon or Papua New Guinea and that the Blackbirds were birds-of-paradise. I have experienced many dawn choruses around the world in my life, but that first one, in my garden in Finglas, is still the best dawn chorus ever.

The last entry of *Purnell's Animal Life* was on page 2,688. It was about the Zorro. I am sure few people have ever heard of a Zorro. It is in fact the name given to several South American dog species that are fox-like but which are not true foxes (*zorro* being the Spanish for fox). Even to this day, this six-volume set is an incredible source of reference and knowledge. It's also hard to believe that the very young Eric Dempsey read each one of those 2,688 pages.

Collecting the last issue of *Purnell's Animal Life* was a very sad moment. I went into Dagge's and, as always, Leonie had my copy ready for me. I knew that I would miss this weekly wildlife assault on my senses. However, as I was leaving the shop, I was stopped in my tracks by something new on the shelf. It was a brand-new weekly series that had just been launched. It was *The Encyclopaedia of Birds*. I couldn't believe it. I went from *Purnell's Animal Life* straight into *The Encyclopaedia of Birds*.

Birds was a revelation as it dealt with birds of Britain and Europe (so many books still refer to 'the birds of Britain and Europe' as if Britain is not part of Europe). However, reading through the pages, I suddenly realised that many of the birds that I was reading about could actually be seen in Ireland. It was this realisation that changed my life utterly. Now I wanted to see the birds around me. I was no longer just looking at birds when I was out: I was actually going out to look at birds. It may seem to be the same thing, but it is radically different. I was now seeking out bird experiences of my own. I was trying to identify the birds I was seeing. I was a fledgling birdwatcher forging a path that I would follow for life.

The Encyclopaedia of Birds ran for a total of seven volumes, the last two of which dealt with exotic birds from around the world. The last entry of Volume 7 is about the Red Bird-of-Paradise, a bird I have yet to see.

Reading these wonderful encyclopaedias created ambitions in my very young mind. We all had childhood ambitions. Perhaps it was to play for your favourite football team or to be a train driver, an astronaut or a film star. Few of these childhood ambitions are ever realised. However, I consider myself very lucky in that many of my childhood ambitions have been fulfilled. Many of the animals and birds that I promised myself I would experience some day, I have experienced. Birds that I remember reading about and that I dreamed about seeing when I was nine years old, I have now seen. I am a lucky man, because many of my childhood ambitions are still being realised. I am a lucky man to have been given a love of nature from such an early age. I am a lucky man to still hold a child-like sense of wonder about the natural world around me, a sense of wonder sparked by Ma's gift of *Purnell's Illustrated Encyclopaedia of Animal Life*. Ultimately, I am a lucky man because, as a child, I was given the greatest gift of all – a love of reading.

As I reflect on this, I wonder what the 53-year-old Eric might write to Santy and ask for?

I think I would ask for a ball and a box of dog chocolates for my beautiful spaniel Suzie.

And for me?

'Well, Santy, if you're listening, please bring me a lot of books.'

Chapter 4 ➔

SHEBA — A CONSCIENTIOUS LEFT-WING GUINEA PIG

Sheba was an Abyssinian. She was lovely. She had beautiful dark eyes and the most perfect straight white teeth. Her golden and ginger hair stuck up all over the place. It reminded me of my own hair when I got out of bed in the morning. I always liked Abyssinians; I never took to smooth-haired guinea pigs.

Sheba was my pet guinea pig and I got her for my eighth birthday. She was a new member of our family, but was one of a large tribe of guinea pigs that lived in Finglas. She was friendly and would allow us all to pick her up and pet her.

I kept her in a cage in the coal shed. The cage had loads of room for her to run around in and had sawdust on its floor to keep her warm and dry. There was also a little hut in the cage where she slept and I put straw in that so she had a nice bed to sleep on each night. She loved to eat fresh grass and dandelion leaves. The coal shed always smelt of freshly cut grass. I used to sneak her lettuce, cabbage, apples and carrots as well.

On fine days I would bring her out to the back garden where I had a large wire run for her. She was safe inside that and would eat the grass down to a fine tip. Da thought this was great because he didn't have to cut the grass. Moving the run around the garden created a 'mosaic of cut-grass circles' as Da called them. I didn't know what he meant by 'mosaic' but I could see all the cut-grass circles very clearly. If it rained, she had a little waterproof shed in the run where she could shelter. There was no doubting it: as far as guinea pigs were concerned, Sheba had the life of Riley.

I fed her every morning before I went to school. She'd hear me coming to the door of the coal shed and would start calling. Her 'weak-weak' calls would start off quietly at first but would then get louder and louder. When she got into full voice she'd scream the house down. Clare nicknamed her 'the weak'.

In the summertime, Sheba joined the herd of free-range guinea pigs that lived with the Redmonds next door. Fred was the leader of this herd. He was a smooth-haired guinea pig but, like Sheba, he was also ginger and gold. These guinea pigs slept in their family coal shed and in the morning, as soon as they heard one of the Redmonds approaching, they all started calling from the darkness of the shed. I could hear them even from my bedroom.

As soon as the coal shed door was opened, they'd all stream out and start eating the grass in the Redmonds' back garden. Their grass was so short Da reckoned you could play snooker on it. Looking out at their garden from my bedroom, I could sometimes count as many as twenty guinea pigs feeding in the garden. I imagined being in Africa and looking out at herds of animals on the grasslands just like I saw on telly. They were constantly calling to each other; all it took was for one of them to start 'weaking', and the whole lot would be off. Sometimes it sounded like a guinea pig dawn chorus.

The Redmonds also had a big dog called Prince. He was a bit like a German Shepherd but he wasn't one. Prince once chased a fella on a motorbike and nearly knocked him off it. The trouble was that the fella he chased was a motorbike guard and Prince got into big trouble. He was brought to *court* and Ma said he had a 'record' and was on his 'last chance'. It was a good job that no one ever found out that he had taken a lump out of the gasman. For a dog that had taken a bite out of the gasman and nearly knocked a motorbike guard off his bike, Prince was a great dog – he never touched a single guinea pig.

When Fred was killed by a moggie in the Dalys' garden (he must have escaped from the Redmonds and found his way into the Dalys' garden, where the cat caught him), we decided to put Sheba back into the safety of her cage and her days of eating grass in the Redmonds' garden were over. I think she really missed not being able to run free.

Johnny Bourne also had a guinea pig. She was mostly white. He called her Snowy, which I thought was a great name. Johnny lived across the road from us, so Snowy wasn't part of the tribe of guinea pigs that lived with the Redmonds. In fact, Snowy rarely met other guinea pigs and Johnny didn't have a run for her either. One fine sunny Sunday morning, I called over to his house. We were going to Mass around in the Tin Church. I was going to the Tin Church because Da was doing overtime in the Mater. If Da wasn't working he used to take us to the 'Woodener' in Glasnevin or to the church in Berkeley Road near where he grew up.

Before heading to Mass, Johnny and I visited Snowy in her cage. I thought she looked a bit sad sitting there by herself. It was such a shame that she was stuck in her cage on such a lovely day. In fact, it was such a shame that we'd be stuck inside the Tin Church on such a lovely day. Thinking about this, I hatched a plan.

'Maybe we can tell our Mas we're going to Mass, but we'll bring the guinea pigs up to the park instead,' I suggested.

It seemed like a great idea and Johnny was up for it. We decided that we would hide our guinea pigs under our jackets and sneak them past our Mas and out of the house, pretending we were going to Mass. Then we'd walk as if we were going to Mass but mitch off and go to the park instead. The guinea pigs would have a ball. They'd be able to run around, eat grass and get to know each other. As well as that, from the park we would be able to see when Mass had finished and then we could hide the guinea pigs again, go home as if we'd been to Mass and put the guinea pigs back into their cages. No one would ever know. It was a fail-safe plan; it was perfect.

With the plan now in place, I headed home and put on my 'bomber' jacket. It was called a bomber jacket because men who flew bombers used to wear them (well, that's what I'd heard anyway). It was perfect for my purposes because it had a zipper at the front and a tight band around the waist. Once I put Sheba inside, she wouldn't be able to escape. When Ma wasn't looking, I sneaked out to the coal shed, took Sheba out of her cage, tucked her into my jacket and zipped it up good and tight. I quickly walked back through the house to the front door, announcing loudly that I was off to Mass with Johnny. Once outside, I walked down to the lane behind the shops and waited there. I had my jacket zipped up enough to stop Sheba from escaping but had to be sure there was enough air getting in for her to breathe. I didn't have to worry, because she was very relaxed and seemed to be going to sleep. A few minutes later, Johnny arrived. He was wearing his bomber jacket too, with Snowy tucked snugly inside. Part one of our plan had gone smoothly.

From the laneway we started to walk past the shops and around the corner by the garage. There were loads of people heading to 11 a.m. Mass, which was very popular. Da used to say that only Holy Joes went to early Mass – everyone else was too lazy on Sunday mornings to get out of bed.

We continued past the garage and approached the church. At this stage most of the people crossed the road, but we didn't. Instead, we hugged the walls of the houses along the left side of the road, opposite the church. We kept our heads down and, as we passed the church doors directly across the road, we quickened our pace, without breaking into a run. Once past this point we started meeting people coming down the hill to Mass. We didn't worry about them because they weren't from our road and didn't know us. Our plan was working. We reached the park and entered through the small gate. We'd made it.

I loved Johnstown Park. There were trees and flowers everywhere. It was a great place to watch butterflies or to catch bees to look at before letting them go. This lovely morning, we sat down on the grass and watched Swallows swooping low over our heads and catching insects. There were young Swallows on the wires and they were sitting calling to be fed. I told Johnny what I had read about Swallows: that they would head all the way to Africa for the winter. I told him that when they came home to Ireland each summer, they could even find the exact nest where they were born. He was amazed but I wasn't sure if he really believed me or not. I didn't really care. Sitting in the park was better than Mass any day. We were just about to release our guinea pigs into this little piece of paradise when, from out of nowhere, came a voice.

'Is that you, Johnny Bourne?'

We were rooted to the spot. We looked around. It was some aul' wan I didn't know. She was small and fat with wrinkled, squinty eyes and blue hair. I didn't like the look of her.

'Eh, hello, Mrs Burke,' Johnny said. He obviously knew her.

'And what are you up to?' she demanded to know.

I stayed quiet.

'Nothin',' replied Johnny, adding, 'We're just goin' playin' in the park.'

She looked at him with her squinty eyes. She didn't look at me at all, which I was happy about.

'I hope you're not meant to be at Mass?' she said.

Jaysus, I thought to myself, she's quick at coppin' on to things!

'No!' Johnny replied. 'We've already been to Mass.'

He lied so well that I almost believed him myself.

'Is that so?' she persisted. 'So, if I tell your Ma that I saw you in the park at eleven o'clock, she'd be all right with that, would she?'

This aul' busybody was good. I could picture her in a war film standing in a prisoner-of-war camp interrogating some poor fella she had captured. I could imagine her in a soldier's uniform with her blue hair sticking out from under her cap. No one could survive her questions and poor aul' Johnny buckled instantly.

'No, Mrs Burke,' Johnny admitted, looking down at the ground, his face going all red. 'But …'

'No "buts",' she interrupted. 'Get down that hill and go to Mass before I tell your Ma.' She looked at me. 'And that goes for you too!'

With that, she marched us out of the park, escorted us down the hill and delivered us to the front doors of the church. Then she stood waiting for us to go in. We pushed the glass and wooden doors open and entered the porch of the church. Once we had gone inside, she turned on her heel and went on her busybody way. We were 'sweating it', but we were okay because we were only inside the porch of the church.

'If we wait till she's gone, we can get out of here,' I whispered.

Through the glass of the doors, we watched as she rounded the corner by the garage. At this stage I could feel Sheba was getting a little bit annoyed at being stuck inside my jacket. And it looked as if Johnny had a big lump of belly that had a life of its own: Snowy was also beginning to have enough of being inside his jacket. We continued to watch Mrs Busybody Burke as she walked up towards the shops and went out of sight. It was time to make our move.

However, just as we opened the doors and tasted our freedom, the inner doors of the church porch swung open. There stood the aul' fella who was in charge of everything to do with the church. I didn't know his name but knew that he always dressed like a priest. He wore a long black gown over his shirt and trousers and was 'all the business', according to Da. In fact, Da said that 'he probably wanted to be a priest but wasn't holy enough!' He looked after the baskets of money at Mass and it was always him that held the big candle on special occasions like the blessing of the shamrocks on St Patrick's Day. He was also the one who rang the big church bell outside to remind us all to say the Angelus. And, on Sundays, he was the one in charge of finding everyone a seat for Mass.

He had heard us whispering inside the porch. He was another aul' busybody and the one thing he hated was young fellas dossing in the porch when they were meant to be at Mass. When I saw him, I knew there was no escape.

'Get in here, lads,' he said. 'I don't want ye messing around.'

He always said 'ye' which, according to Da, meant he was a culchie.

'But we're not going to Mass,' I said. 'We can't!'

'Get in here, ye, and none of your lip,' he answered as he ushered us into the church.

Johnny and I looked at each other. We needed to come up with a plan very quickly. But the best I could come up with was simply to repeat 'But we can't.' He stopped and looked at me.

'And why not?'

I didn't have an answer for him. I could hardly tell him that we had very annoyed guinea pigs inside our jackets. I could only mutter, 'We just can't!'

He was having none of it and pushed us up the centre aisle of the church. I spotted a few spaces near the side doors and offered to sit over there. At least we could slip out of Mass during Communion. He was on to my plan straight away.

'Oh, no,' he said. 'I'm not having ye slipping out of Mass during Communion.'

He brought us up past the middle of the church and delivered us to a row of people right up at the front of the church. We 'excused' our way past people and found our seats in the middle of the row. Mass had just begun and I looked up to see Father Farrelly on the altar.

'It's Farrelly,' I whispered to Johnny.

Father Farrelly was an aul' fella. He walked really slowly around the altar and spoke even slower. When he said Mass, all the prayers took ages. Everyone would be finished their Hail Marys and their Our Fathers long before he was and he'd be left saying them by himself. And when he stood up to deliver his sermon on the pulpit, he took for ever because he spluttered and coughed as he spoke. Sometimes he even forgot what he was talking about. Da once said that even the Holy Joes looked at their watches when Father Farrelly was saying Mass. I knew in my heart that this was going to be the longest Mass ever.

As I sat down, I zipped my jacket right up to the neck. I could feel that Sheba was getting a bit hot and angry in there and she was moving around. I knelt down and joined in the Mass, saying my prayers. I hoped no one would notice the lump that was moving around inside my jacket. Looking at Johnny, I could see the lump of Snowy moving around inside his. Beads of sweat broke out on my forehead.

We prayed our way through the first part of Mass and Sheba seemed to be settling down a bit. I thought that maybe all the walking had annoyed her but that she was getting sleepy now that I was just kneeling, standing and sitting. I unzipped my jacket a little and peaked inside to check on her. She stared back at me with wide-open eyes. She certainly wasn't getting sleepy.

Father Farrelly finished some prayers and strolled slowly across the altar, stopping to bow in the middle, before making his way into the pulpit. He stood up in the pulpit and turned on the microphone. He even turned the microphone on slowly. He then coughed and started blowing his nose. He always blew his nose during Mass.

Why didn't he blow his nose before saying Mass? I wondered. Does he like the sound it makes in the microphone?

'This week, I'd like to talk about the evils of the trade union movement and socialism,' he began.

Da said this was Father Farrelly's hobby-horse. I remembered once being with Da when Father Farrelly started going on about it. There was a strike on somewhere and he started telling everyone at Mass that it was their duty to pass the pickets.

'Trade unionists and socialists are doing the work of the Devil,' he said. 'They want the evils of communism to descend on Holy Ireland. As a good Roman Catholic, it is your duty never to go on strike. Know that God is on your side when you pass the pickets of these evil people ...'

With that, Da nudged me. 'Come on,' he said. 'We're going.'

Da got up and I followed. We 'excused' our way out of our seats and Da marched down the church as Father Farrelly stood silently with his jaw open. I was holding Da's hand. People were looking at us. Da kept his head up and eyes straight ahead. He walked strongly down the centre aisle with me beside him. I felt a surge of great childish pride that I was walking out of Mass with Da. I didn't understand why, but I knew that whatever aul' Farrelly had said had really annoyed Da.

When we got outside, Da smiled at me. He knew I hadn't a clue why we had just walked out of Mass.

'Some day when you're old enough, you'll understand,' he said.

So aul' Farrelly was once again on his hobby-horse and I knew this Mass was going to go on and on. As he was about to speak his next sentence, my world fell apart. Sheba had had enough of being inside my jacket and decided to let me know.

'Weak!' she uttered quietly.

Father Farrelly stopped briefly. A few people close by looked around to see where the sound had come from. I decided to put a puzzled look on my face and started looking around as well. I even shrugged my shoulders at the aul' wan beside me. Johnny kept his eyes closed and seemed to be praying as if his life depended on it. In the pulpit, Father Farrelly coughed and began to speak again.

'Socialism is the root of evil in this world. It is against the work of God and all …'

'Weak!'

This time Sheba protested much louder than before. People looked around again but this time they were looking at me. I coughed loudly hoping that they might think I just had a frog in my throat. Father Farrelly had stopped speaking and was now looking around from the pulpit with a very puzzled look on his face. I kept my eyes facing forward, but I knew my face had gone red.

Father Farrelly coughed and blew his nose again before starting from the beginning.

'Socialism is …'

'Weak-weak-weak-weak!'

This time Sheba had had enough. She was weaking as loud as she could. And within seconds Snowy had joined in with her. The pair of them were weaking as if they'd known each other for years.

'Weak-weak-weak-weak-weak!'

'What is going on?' Father Farrelly demanded to know from the pulpit.

It was the first time I ever heard him speaking quickly and without coughin' or splutterin'. Shame he didn't speak like that all the time.

'Weak-weak-weak-weak' came the answer as Sheba and Snowy took over the Mass.

By now everyone was staring at us and an aul' wan beside Johnny pointed us out to Father Farrelly as he once again demanded to know what was going on. By this time the weaking of Sheba and Snowy had reached its highest level. Their calls echoed off the walls and the roof. It sounded like there were hundreds of guinea pigs in the church. The sound was magical but I didn't get much of a chance to appreciate it.

'Get out of this church,' Father Farrelly thundered at us from the pulpit, his face red with anger.

'Escort those bowsies out of God's house,' he shouted to the aul' fella who was in charge of everything.

He was beside our row of seats in seconds. We were forced up from our seats and excused our way out into the centre aisle. All the time, Sheba and Snowy weaked to their hearts' content. Once we were in the aisle, the aul' fella who was in charge of everything grabbed us by the scruff of our jackets and pulled us down the church. Everyone was frowning at us and giving us nasty looks. With each stride, the echoes from the guinea pigs' weaking seemed to change. They seemed to be bouncing off the end wall of the church. It was better than any weaking I'd ever heard before. I wondered what it would sound like if all of the Redmonds' guinea pigs were here as well! As I thought about that, I found myself grinning. I glanced over at Johnny beside me. His face was bright red and he had tears coming down his cheeks. I knew my face was red too because it felt really hot. But I wasn't crying. No, for some reason, my grin was turning into a titter and by the time I was in the middle of the church I was beginning to laugh.

'Get those bowsies out of here!' shouted Father Farrelly as we cut across an aisle leading to the side door. 'Your families should be ashamed of you.'

The aul' fella who was in charge of everything saw me laughing.

'Do ye think this is a laughing matter?' he asked, but it was Sheba who answered him.

'Weak-weak-weak-weak!' she replied, and Snowy agreed.

When he got us into the porch of the side door, out of view of everyone, he pushed us down the steps and out of the church. He also gave me a kick in the arse. He didn't kick Johnny, because he was crying. I think I got the kick because I was laughing. It didn't hurt. Once outside, we decided to go to the park and recover from what had happened. I was shaking but still laughing. Poor Johnny was in bits.

'Me Ma is goin' to kill me,' he stated.

'Mine too,' I admitted. 'But did you see the face of Father Farrelly when Sheba started "weaking"? The head on that and the price of cabbage, eh?'

Johnny looked at me and his face broke into a grin, which very quickly turned into a laugh. We both just sat there with tears coming down our faces as we laughed. Our tummies were sore from laughing.

'Jaysus, I thought he was goin' to have a fit,' Johnny replied when he'd stopped laughing.

We decided we'd stick with the plan and we let Sheba and Snowy have a good run around the park. We felt that they earned it. When we saw people starting to come out of Mass, we picked them up, put them back inside our jackets and began walking home. Both guinea pigs were as quiet as mice. We took a long way home, down through the muddy path along the top of the park. That way we would meet fewer people who might recognise us. As we got closer to home, we saw an aul' wan talking to Johnny's

Ma. She was obviously telling his Ma what had happened, because she was wagging her finger and shaking her head. This was our third aul' busybody in one day. It seemed there were busybodies everywhere.

As soon as Mrs Bourne clapped eyes on the both us, she told Johnny to 'get inside at once'. He was 'goin' to be killed!'

I went home and slipped in past the kitchen, where Ma was making the dinner, and out the back door to the coal shed, where I put Sheba back in her cage. I then sat on the step at the back door and thought about what had happened. I just couldn't help grinning when I thought of aul' Father Farrelly's face as Sheba starting weaking. But the grin was wiped off my face when I heard a knock on the front door. Busybody number four had arrived in the form of Mrs Smith from up the road. I could hear her telling Ma all about it. In fact, she told it very well. It was a great story. The talking stopped and the front door closed. I heard Ma walking down the hall and towards the back door. Her steps were heavy on the lino. The back door opened behind me.

'Get in here, you,' she said.

I followed her into the kitchen and prepared to face what was coming to me.

'Is it true what I've just been told about you?' she asked.

'Yes, Ma,' I replied very quietly.

'What on earth possessed you to bring Sheba to Mass?'

Before I could begin to tell her, she told me that I'd made a holy show of her and that she didn't want to hear another word from me.

She sounded really angry, but as she turned away from me I thought I could see a grin on her face.

'Just wait till your father gets home later,' she added as she walked away. 'That's all I have to say!'

When Da came home that evening, I was upstairs in my bedroom reading my latest issue of *Purnell's Animal Life*. I could hear Ma and

Da talking downstairs and a few minutes later I heard Da calling me.

'Eric, get down here,' he said. 'I want to talk to you.'

I took a deep breath and came down to the kitchen where Ma and Da were standing. Da looked annoyed and stared at me.

'So?' he said. 'Tell me what happened today.'

I decided to confess to the whole thing. I told him about feeling sorry for Snowy stuck in a cage; how we planned to mitch off Mass and bring the guinea pigs up to the park; and how we had put them inside our jackets. I told him that we met an aul' busybody who forced us to go to Mass, but we were only in the porch when the aul' fella who was in charge of everything in the church found us and made us go inside. And then I told him that Sheba started weaking in the middle of Mass and that Snowy had joined in. With my head down, I told him that we'd been thrown out of Mass.

'And who was saying Mass?' Da asked.

'Father Farrelly,' I replied.

'Farrelly, eh?'

'Yes, Da.'

'And when did Sheba start her weaking?'

'Just when Father Farrelly was about to start talkin' from the stand,' I answered.

'She started weaking just when Farrelly was about to give his sermon from the pulpit, eh? And can you remember what his sermon was about?'

'Yes, Da. It was about that stuff you say is his hobby-horse.'

'Was he goin' on about the evils of socialism and the trade unions? Is that what you mean?'

'Yes, Da, that's the stuff he was startin' to talk about until Sheba started to weak.'

A broad smile swept across his face.

'Well, now, I can't say I blame Sheba for that,' he said. 'I often feel like screaming meself when that aul' fella starts preachin' about that.'

He smiled again before pouring himself a cup of tea, picking up the newspaper and sitting down at the table.

'So she started weaking before aul' Farrelly could start talkin' about the evils of trade unions and socialism and the like?' he said, more to himself than to me.

He put his paper down and smiled at me.

'I would've loved to have seen his face!' he said.

Then, with an even broader smile, he added …

'Well, now, isn't it great to know that we have a little conscientious left-wing guinea pig living in our midst?'

Chapter 5 ✦

LOST INNOCENCE

When it came to the summer holidays, Da was always full of great ideas. He was full of 'enthusiasm', as Ma called it. Each year, he took his two weeks' holidays and he'd bring us to the seaside, on tours of museums and galleries, and even on 'mystery train tours' organised by CIÉ. Not only that, but he'd also bring our friends along to keep us company. Ma didn't go with us that much.

We'd get up early, pack our lunches and head off with our bags on the bus into town. Then we'd either catch the bus to the seaside or walk to the museums and galleries. When we went on the mystery tours, we'd walk down Talbot Street to the train station and Da would get our tickets. We'd get on the train, trying to guess where we might be heading.

Some years, Da would even hire a car. We never had a car of our own so it was great to have one for a week in the summer. We'd go off each day to different places. Ma came with us once. We went up into the mountains but Ma got bored very quickly. Poor Da was killed trying to show us different sites, but after about an hour Ma wanted to go home.

'You've seen one mountain, you've seen them all!' she said.

Ma didn't come on too many trips after that. However, our excitement at travelling off each day in a car was never dampened. Eventually Clare was getting too old to want to be seen going anywhere with Da so, in July 1973, he announced that this would be the last summer he'd hire a car. He took Johnny Bourne and me to loads of places during the week we had it. As the end of the week came close, Da said that he was bringing Gerald out for a drive on Saturday. He asked me if I wanted to go with them.

Uncle Gerald was Da's older brother by about twelve years. My second name is Gerald and I was named after him. I didn't get to meet Uncle Gerald that often. I'd see him maybe a few times each year, and he always spent St Stephen's Day with us. He lived close to town near the Black Church, in a tenement building in Mountjoy Street which was also where Da was born and raised. The house was dark and run down, but Uncle Gerald still lived there in one tiny room.

Whenever I visited Uncle Gerald, I would sit in his room and look around me, fascinated. The room was a bit dirty, to say the least. There was a narrow metal bed in the corner, a table and two wooden chairs, a small fireplace and a gas cooker that was covered in grease. A tiny window looked out onto the street below, but you couldn't see much because the net curtains were dirty yellow and the window was stained and cracked. The room was dark and there were cobwebs hanging everywhere. Uncle Gerald smoked and there was always an ashtray full of ash and cigarette butts. The room smelt of smoke and old cooking. Among this general untidiness lay piles of books stacked on the floor and on the window ledge. As a kid I used to look at them. They were about the politics and history of Russia and Ireland, as well as big, thick, heavy novels. However, amongst them all were loads of poetry books. These were all the same kinds of books that Da had, but his were placed neatly on shelves in his room.

Uncle Gerald was a bald, rotund man with a happy, red-cheeked face. He wore old clothes and never used a belt to keep his trousers up; instead, he used a tie fastened around his waist. His smile was big, wide and gummy. He hadn't a tooth in his head and never bothered getting false teeth. Da told me that he never worked but spent all his time writing poetry or reading books. He was what Da called 'bohemian'. Da used to get annoyed with him because, when he'd write a new poem, he would often roll it up, stick it in the fire and use it to light a cigarette. Da said that Uncle Gerald didn't really appreciate the fact that he had written a poem. It was as if once he had it out of him, it didn't matter any more. Da used to try to rescue and keep the poems Uncle Gerald wrote. I still have many of them, written on scraps of paper in Gerald's handwriting.

To me, Uncle Gerald was always someone to be around. He spoke quietly, gently and slowly, in a real old Dublin accent. It was as if he thought about each word before speaking it. I always felt that each word he said was worth listening to. I thought the same about Da. They were great friends and met each other every Saturday night for a pint and a chat (or a major political discussion). Da called in to see him every day for a cup of tea because Da worked in the Mater Hospital, which was only up the road from where Uncle Gerald lived. So when Da asked if I'd like to spend the day out with them both, I didn't have to think about it for a second. I wouldn't have missed it for the world.

That Saturday morning was warm and sunny. Da and I had a quick breakfast and set off in the orange Mini that Da had hired. We didn't pack a lunch because Da said we'd be eating lunch in a pub. We drove to Mountjoy Street and parked the car outside number 58. The front door was open and we walked in and up the narrow, dirty and dark stairs. Uncle Gerald's room was the first on the left off the first landing. We went in. He was up and having a cup of tea and a cigarette. He welcomed me with a firm handshake. His hands were smooth and he had long fingers. Ma used to say he had great fingers for playing the piano. He poured us a cup of tea and I sat listening to Da and him talking.

'So, Tommy, where are you taking us?' he asked.

Da smiled but wouldn't tell him. He was bringing us on his own mystery tour. We packed into the car – me in the back; them in the front – and we headed out of Dublin, up past Glasnevin Cemetery and Finglas. I hadn't a clue where we were going and I didn't care. I was happy to be with Da and Uncle Gerald, sitting back and listening to what they were talking about. They talked about everything, but they talked a lot about poetry and writers. About forty minutes into our journey, Da decided to tell us where we were going.

'We're going into Ledwidge country,' he announced.

I didn't know what that meant but the news was greeted with a big toothless smile from Uncle Gerald. It turned out that both he and Da loved a poet called Francis Ledwidge. He was killed in World War One. Uncle Gerald explained that he was one of the war poets.

'You'd like him because he is sometimes called "the poet of the Blackbirds",' he said.

I liked the sound of that.

We drove along narrow country roads, and I looked out at the trees and the bushes with white flowers. We then arrived at a little cottage. This was where the poet had lived. We all got out and walked around. It was lovely. Birds were singing everywhere. A Robin was in full song in the hedges close by and Swallows swooped for insects over the nearby fields. Da and Uncle Gerald spent a lot of time walking around the cottage. Da looked at the old trees in the fields beside the house.

'Think about it,' he said. 'It's possible that Francis Ledwidge sat under those trees. If only they could talk … I'm sure they could tell a great story.'

As a kid, I was always fascinated when Da said things like that. My young mind could imagine the trees talking and telling me all the things they had seen in their lives. I wondered what they would tell me.

'Ah, indeed,' Gerald answered. 'Who knows, perhaps "He shall not hear the bittern cry" was written under those very trees!'

With that, the two of them started reciting:

He shall not hear the bittern cry
In the wild sky, where he is lain,
Nor voices of the sweeter birds
Above the wailing of the rain.

It was a sad poem, but also a very lovely poem. I stood back
and listened to them enjoying saying it out loud together. There
was no doubting that Da and Uncle Gerald loved poetry. There
was no doubting that Da and Uncle Gerald were great friends. I
wondered what it would be like not to be able to hear birds. I had
heard of a Bittern. I wracked my brain to remember more about
them. As far as I could recall, they looked a bit like a heron but I
just couldn't be sure if they were in Ireland. I vowed to check that
when we got back.

When we were finished at the cottage, Da drove to a small pub
in the middle of nowhere. He ordered some sandwiches and tea,
and we all sat on a wooden seat outside, enjoying the warmth of
the sun and the blue sky overhead.

During the whole day Da and Uncle Gerald talked to each other
and to me. They always made me feel that what I had to say was
important. They would listen to me and ask me for my opinion
on things. But most times I was just happy to listen. They never
seemed to run out of things to say to each other and they always
sounded really wise.

After we finished our sandwiches, Da drove us to the ancient
burial site of Newgrange. There was no one else there. We walked
around the mound, touching the stones that surrounded it in a big
circle.

'Hard to believe that these stones were placed right here by
people who lived thousands of years ago,' Da remarked. 'I wonder
what Ireland looked like back then?'

Da and I even went into the tomb. It was cold and dark inside.
Uncle Gerald didn't go in. He said he was too fat and he might get
stuck. We all laughed at the idea of him stuck in the middle of the
tomb.

From here, we continued to drive around the countryside, and by late afternoon my young mind was filled with poetry and history. I was exhausted.

Da started driving back to Dublin. Ma said she'd have a dinner ready for us and Uncle Gerald was coming to our house for dinner. When we got home, we sat around the kitchen table and I told Ma all about Francis Ledwidge and the tomb at Newgrange. Clare sat down and joined us. She had been out with her pals all day. Paul was upstairs playing his guitar. He had waited to see Uncle Gerald before going out with his pals for the evening. When he came downstairs, he and Uncle Gerald greeted each other like they were old friends. Uncle Gerald used to look after Paul when he was a baby.

'You both missed a great day out,' I told them.

'I doubt that you'll be wanting to go out with your aul' Da for much longer,' Da said, smiling at me. 'Sure you're twelve years old next Tuesday.'

'Yes,' agreed Uncle Gerald. 'The days of lost innocence are approaching fast!'

I didn't really know what that meant, but it did sound very wise.

After dinner, Da and Uncle Gerald sat talking. Da poured them each a small glass of whiskey. Da never drank when he was at home and only drank whiskey on special occasions. It seemed that this was a special occasion. They sat talking for hours. I sat with them and added my tuppence-ha'penny worth. They listened to what I had to say but didn't always agree with me. They even argued against some of the things I said.

However, they couldn't argue with my bird facts. As soon as we'd returned home that evening, I had raced upstairs to consult my *Purnell's Illustrated Encyclopaedia of Animal Life*. There, on page 219 of Volume One, was all I needed to know:

Bitterns are members of the heron family and have very cryptic patterned plumage (almost owl-like). They frequent large areas of reed beds and, when alarmed, tend to freeze, and raise their heads towards the sky and remain still, their plumage providing ideal camouflage. The song of the male is a deep booming, usually heard at night.

However, reading on, I discovered that the species was very rare in Britain due to overshooting and the extensive drainage of marshlands. They were now only found in very small numbers in the wetlands of Suffolk, on the east coast of England.

Further research revealed that the truth was even sadder from an Irish perspective. It appeared that the birds were once a very common species in most parts of Ireland, but, as happened in Britain, overshooting and habitat loss resulted in Bitterns becoming extinct as a breeding species in Ireland by 1840. Now, they were only a rare winter visitor when cold weather in Europe and Britain forced them westwards to Ireland in search of open, unfrozen water where they could fish or catch frogs.

Armed with all this information, I advised Da and Uncle Gerald that, having checked my books on birds, I had to let them know that 'your man' Francis Ledwidge might have been a great poet, but he didn't know his birds very well. If that poem *was* about one of the men shot after the Easter Rising in 1916, which Da and Uncle Gerald said it was, then he would never have heard a Bittern cry … ever! Even if he lived to be 100 years old. They were gone out of Ireland by the middle of the 1800s.

They acknowledged my superior 'ornithological prowess', as Uncle Gerald called it, but still would not allow a minor fact such as this to detract from the beauty of the poem. For me, it felt wonderful to be sitting there talking with them. I felt grown-up, but by 10.00 p.m. I was getting really exhausted. In fact, when I yawned, Uncle Gerald yawned too. It was time to call it a day, as Da often said.

Uncle Gerald said goodnight to Ma before Da brought him home. I went along to keep Da company. It was a lovely warm night. We drove towards Mountjoy Street, heading down through Glasnevin. There were lots of people out walking. Going down

Washerwoman's Hill, Da had to swerve the car into the middle of the road to avoid a big gang of fellas who were standing in the road. They were shouting at people and drinking from bottles of beer. They were across from the Tolka House.

'Now there's a right gang of skinheads and boot boys,' Da said. 'Right thugs.'

Uncle Gerald agreed, adding that they looked like a right 'gang of bowsies'.

We dropped Uncle Gerald home and we watched him go inside the house. I got into the front seat and sat beside Da. He took the long way home and, by the time we got back, I was almost falling asleep beside him. I thanked Da for bringing me out with him and Uncle Gerald and went straight up to bed. I fell asleep thinking of poetry and Francis Ledwidge fighting in the war. I was thinking about ancient tombs, about what the trees would tell me if they could talk and about Bitterns. I promised myself that I would see a Bittern one day. My thoughts and dreams became one as I fell into a deep sleep.

I didn't know what time it was when I became aware of Da in the room. I wasn't even sure if I was dreaming or not. It seemed bright so it must have been morning, but it wasn't sunny so maybe it was really early. In my semi-conscious state, I could hear Da talking really quietly.

'Are you okay? Here, let me help you with that,' he was saying.

Then I could hear Paul. We shared a bedroom: I was on the left side with my books and magazines, and he ruled the right side with his record player, LPs and guitar. He also had a big speaker that he plugged his guitar into. When he was playing his guitar, the whole house would shake. But Paul wasn't saying much back to Da.

'Yeh, I'm fine,' was all he said.

Da's voice sounded worried.

'Are you sure? Here, let me help you take that off,' Da said.

'Thanks, Da,' Paul said.

As I became more awake, I also became more aware of the meaning of the words being spoken. Da was helping Paul to get into bed.

'Ah,' I thought to myself. 'Paul must be a bit drunk.'

Paul had been out with some of his friends so maybe he'd a bit too much and Da was helping him into bed. I had my face to the wall, so I couldn't see them.

'Sit down there and I'll help you off with your jeans,' Da said.

As Paul sat on the bed, it creaked a little and he let out a long, deep groan. I could hear Da bending down and pulling Paul's boots off, one and then the other falling on the floor with a dull thud. Da continued to talk really quietly.

'Sit up there for a second and we'll get those off you.'

'Jaysus,' I thought. 'Paul must be really drunk if Da is helping him out of his clothes.'

With the quiet way Da was talking, it was obvious that he didn't want Ma to wake up and find out. I thought Da was great to be doing this. I could hear the effort it took to get Paul's jeans off him. Da put them on the back of the chair. I heard the buckle of the belt hitting the wooden arms.

'Right,' said Da. 'Let's get that T-shirt off you and get you into bed.'

And then I heard a sound that made me wake up fully in an instant. It was a sound that I didn't expect. It was a long gasp of pain.

'Ahhhhh!'

'I know, I know,' Da said really gently. 'I know it must hurt, but it's better if we get it off.'

Paul gasped again as Da went to take his T-shirt off him. I turned around in my bed and looked over. Paul was sitting on his bed and Da was standing over him, slowly and gently removing Paul's T-shirt. It was soaked in blood. I stared as Da slowly pulled it up and over Paul's head. Paul groaned as it left his body and I gasped at what I saw.

Paul's face was huge. It was swollen. It was black and blue. He had a big bandage around his forehead that went around the back of his head. It had stains of blood on it. His eyes were swollen and puffy. He had one big cut that ran down his face from his left eye. I stared and could see loads of threads sticking up along the entire cut – it was all stitched up. His cheeks were like black balloons. His mouth was all swollen and bloody.

Paul saw me looking.

'How'ya?' he said.

It took me a few seconds to even answer him.

'What happened ya? Are ya all right?'

'Yeh, I'm fine,' Paul answered.

Before I could ask another question, Da told me to go back to sleep and that Paul would answer all my questions in the morning. I just sat up in the bed and stared. I watched as Da pulled back the sheets on Paul's bed and fluffed up his pillows. He helped Paul into bed. As Paul was getting in I saw that he had big bandages all over his shoulders and on his back. As he lay down, he sighed. His head was obviously sore, but he seemed to find a comfortable position to lie in. As he pulled the sheets over him I saw that he had loads of cuts and bruises all up his arms and had more bloody bandages on his wrists.

'You'll be okay,' Da said to him. 'Get a good night's sleep. Call me if you need anything.'

Then Da tucked him up like he was small child. He placed his hand gently on Paul's shoulder. Paul reached over and touched his hand. It was a soft touch and they looked at each other for what seemed like five minutes. Paul smiled.

'I'm fine, Da. Really I am,' he said. 'Thanks.'

With that he closed his eyes and almost instantly went to sleep. I continued to sit up in bed, staring over at him. Again, before he left the room, Da told me to go back to sleep. He looked tired.

I closed my eyes and tried to get back to sleep, but I couldn't. All I could do was lie and stare over at Paul in the bed across from me. His bruised, bandaged and swollen face was all I could think about.

I just had to know what happened to him. I knew that one of his friends had a Mini so maybe they'd been in a car crash. His other friend had a motorbike so maybe he'd fallen off that. I just had to know what had happened.

I could hear Da downstairs. He was still up. I got up out of bed very quietly and went down to the kitchen. Da was sitting there by himself. He had a pot of tea on the gas and was drinking a cup of tea and having a cigarette. He looked up when I came in. His blue eyes looked tired and I thought that maybe he had been crying.

'Are ya all right, Da?' I asked him.

'Yes,' he replied. 'But it's been a long night.'

'Was Paul in a crash?' I asked.

'I wish it was a crash,' he replied.

His answer threw me a bit. If it wasn't a crash, what had happened to Paul?

Da answered the question before I asked it.

'It was boot boys and skinheads,' he said with a deep sigh.

I didn't understand what he meant. He saw that I had a puzzled look on my face.

'Boot boys and skinheads gave Paul a good hiding,' he said, looking down and shaking his head.

I sat down and Da told me what he knew. Paul and some of his friends were walking home from the Tolka House when a big gang attacked them. They caught Paul and gave him a good hiding. He was lucky that some man came to his rescue after the gang had left him lying on the ground. He had been attacked in the Cremore area of Glasnevin, on a dark and quiet tree-lined road. The man had taken Paul to the Mater in his car. Da also said Paul was very lucky because he had the best doctor in the Mater looking after him. He'd needed nearly fifty stitches and the doctor told Da that Paul was lucky not to have lost an eye. Finally, Da told me he'd gone up to the Garda station in Finglas about the attack, but the Guards didn't even send out a car to look for the boot boys and skinheads.

'They couldn't have cared less,' Da said. 'I tell you, if I'd have got a hold of them fellas …'

Da said no more. He sounded both sad and angry at the same time. He simply shook his head and took a sup of tea. He poured me out a cup of tea and we sat silently together drinking and listening to the birds singing outside. The dawn chorus was announcing a new day, but even the birds' songs sounded sad.

I went up to bed after a while and tried to get back to sleep. But all I could do was look over at Paul. I closed my eyes and tried to think about Francis Ledwidge and his poetry. I tried to think about the trees around his cottage and the stories they might tell me if they could talk. I tried to think about the Swallows we had seen and about why Bitterns weren't in Ireland any more. I even tried to think about Uncle Gerald getting stuck in the tomb at Newgrange. But all I could think about was Paul getting a right good hiding. Tears streamed down my face.

The following morning, I woke up early. I didn't remember going back to sleep and for a moment I wondered if what I had seen the night before was just a bad dream. I looked over at Paul. He was still asleep. His face was even more swollen and bruised than before. I got out of bed very quietly and got dressed without disturbing him. In the kitchen downstairs, Da, Ma and Clare were sitting having breakfast without saying a word.

'Is he still asleep?' Ma asked me.

I told them he was out for the count. Ma had seen him when he got home from the hospital last night, but Clare hadn't seen him yet. Ma and Clare were crying. I didn't cry because I had cried lots during the night thinking about what had happened.

A little while later, Da took some tea and toast up to Paul. He also brought up clean clothes and brought down Paul's ruined ones. Ma examined them. There were stab holes in his jacket, shirt and T-shirt. They were covered in blood. Ma threw all of the clothes into a bag and put them in the bin. A little while later, Da and Paul came downstairs. Clare got a right shock when she saw him. Her eyes opened wide and her jaw dropped. Paul smiled at her.

It was a warm sunny day but Paul didn't really go out because the sun hurt his eyes. Lots of his friends called over and they spent most of the day upstairs playing records. Neighbours also dropped in to make sure that Paul was okay. They were also making sure that Ma and Da were okay.

I went down to Cremore that afternoon with Johnny Bourne. We walked slowly along the road until we came to a spot where a large pool of dried blood stained the footpath and the road. It was hard to take in the fact that this stain was the blood of my big brother. I felt sick inside.

After dinner that evening, Paul went back upstairs to listen to his records again. He was very quiet all day. Then, at about half nine, he came down.

'Da, will ya drop me down to the Tolka House?' he asked. 'I'm going for a pint.'

Ma wasn't happy and urged him not to go out for another few days until the cuts had healed a little. But Da said nothing. He helped Paul put on his jacket, took the keys of the hired car and went out the door with Paul. About ten minutes later, Da came back alone.

We all sat watching telly, but no one was really watching it. I could see that Ma and Da were really worried about Paul. I was also very worried but I pretended not to be. Clare was also very quiet. It was getting dark outside. Then, just before 11 p.m., we heard the key turning in the front door and Paul walked in.

'What were you thinking?' Ma asked him. 'Da could have gone down in the car and picked you up. How did you get home? Did you get the bus?'

'No, Ma, I walked home by meself!' Paul answered.

We were silenced by his words.

'And I went up through Cremore too,' he added before turning to go upstairs to bed. 'Goodnight, and thanks for lookin' after me.'

Chapter 6

THE WATCHERS

I loved Friday afternoons and the anticipation of being off school for the whole weekend ahead. I was never great at doing 'me eccer' on Fridays. Some of the fellas used to do their homework as soon as they got home, but, having spent a whole week at school, the last thing I wanted to do was to take out books and copies and do homework. No, as soon as I got home, I threw my schoolbag in my room and I was out the door; and I would stay outdoors, roaming the fields, looking for birds and dossing all weekend until Sunday night. Then I'd cram 'me eccer' into a few hours and get it done.

But this Friday was different. It was the last few weeks of being in 6th class. We had done our entrance exams for secondary school and today we were to be informed as to whether we had passed them or not. We were all leaving the Sacred Heart and going off to secondary school. Most of us were heading to St Kevin's College, a series of prefab buildings near the Tin Church and just around the corner from the Sacred Heart. Some of the fellas were going to posh schools. But one way or the other, these would be the last few weeks we'd be all together as a class. It was weird to think that our class, who had known each other since low babies, would be splitting up. Things were about to change for us all. Life was changing. This summer I'd be thirteen years old and would be a teenager at last.

Clarkie, our teacher, had the results for the future St Kevin's lads. I had passed. Ma and Da would receive a letter with booklists and to let them know what class I would be assigned to. I was delighted with myself. I hated exams but didn't find the entrance exam too hard. Then Clarkie announced the best news ever.

'Because you've all worked so hard these past few weeks, I'm lettin' you all off homework this weekend!'

It couldn't get better than that – a whole weekend of freedom without the thought of having to do homework on Sunday. I looked out the window. The weather was nice. We were on the top floor of the school and there was a terrific view over the fields and down into the city centre. Of course, because I liked to look out the window and watch birds going by all day, I was put in the middle of the class where my view was restricted. Clarkie was a bit of a spoilsport. Looking out the window now, I began to plan my weekend's wanderings. This was going to be great.

Clarkie suggested that any of us going to St Kevin's should attend their sports day, which was taking place that day in the fields behind the school.

'It would be good to know what it feels like to be around the older lads in secondary school,' he suggested.

I thought that sounded like a very good idea. So, when we got off at half two, I dashed home, left my bag in my room, gobbled down a lovely banana sandwich and a glass of milk, and made my way around to the fields behind St Kevin's.

The sports day was already well under way when I got there, but I met with some of my pals and we watched the sprinting and the long jump. It was strange to think that next September we'd be part of this gang of fellas. It would be like starting all over again.

It would be like being back in low babies again. Some of the older lads looked like they could have been teachers. Some of them even had beards and moustaches.

I bumped into Johnny Bourne. He was a year ahead of me, so he was already in St Kevin's. He had run in two races but had got nowhere. We were both crap at sports. So I hung around with him for a while. This was great because he started to point out some of the teachers and brothers. First of all there was 'the Duck', the head brother.

'He's a sneaky bollix,' Johnny said. 'You never know what he's goin' to do. If he's lookin' at ya, you're all right. But when he turns away from ya, that's when you're in trouble, 'cause that's when he might throw a punch.'

I took mental notes as Johnny pointed out a few more brothers and teachers that I should steer clear of. Johnny was doing his duty as a good friend, making sure I would be well informed and prepared come September.

Having watched the final race, I was about to head off, but Johnny stopped me. He told me that at the end of the sports day there was a GAA football match between the teachers and the 6th-year pupils.

'It's the only chance some of the older fellas get to kick a teacher,' he added.

That seemed too good to miss, so I stuck around to watch this match along with every other fella in St Kevin's. We were all cheering for the 6th years. Some of the teachers were right little balls of lard and the 6th years were running rings around them within minutes. However, the Duck was the referee and he was giving every decision to the teachers. We all booed whenever he blew his whistle and awarded a free kick to the teachers. By half time, the teachers were up by a point.

The second half started and once again we cheered on the 6th years. Some fellas were even encouraging their mates to give certain teachers a kick.

'Kick the fuckin' legs off Smitty,' one fella close to me shouted.

The Duck instantly blew his whistle, turned around, walked over to this fella and gave him a long hard look.

'There'll be no cursing in front of me, Kelly … do you hear me?' he said angrily, before blowing the whistle to restart the game. He awarded the teachers a free kick.

Johnny nudged me. 'See … told ya he was a bollix!'

I didn't need convincing.

As the game was nearing the final whistle, it was neck and neck. But the teachers were getting very tired and the 6th years were great footballers. One of them picked up a high ball from mid-air and began running the length of the field, soloing as he went (I could never get the hang of running and kicking the ball up and down into my hands). Going at full speed, he punched the ball over the head of a teacher and straight into the hands of one of his team-mates, who started running straight towards the goals. There was not one teacher who could catch him. It was going to be an easy goal; there was only the goalkeeper to beat. We were all cheering. Then it happened.

There was an enormous boom. I had felt it in the ground under my feet. Birds flew from the trees. A Blackbird that had been singing away over the noise of cheering fellas was now silent. It was a sound like no other I had heard before. It was deep and muffled, but it felt huge.

The entire field of shouting fellas were silenced in an instant. The fella who was about to score the goal stopped dead in his tracks and dropped the ball. Everyone just stood silently. You could hear a pin drop. No one moved.

'It must be a feckin' gas explosion!' one fella said nearby. 'And it sounds very close …'

'Jaysus, maybe it's the science room,' said another.

The game of football was abandoned immediately and I joined everyone as we rushed to see if we could spot where the gas explosion was. St Kevin's wasn't on fire. So I headed back to the yard of the Sacred Heart and ran to the steps from where you had a view of the whole area. We all expected to see a house on fire. But there was no house on fire.

Standing on the steps, I gazed out over the fields and towards the city. I could see something in the city: a big thick black plume of smoke rising into the sky. It was rolling in on itself as it climbed higher. Just to the left of it, and slightly in front of it, was another, thinner, plume of black smoke. This second one reached higher into the sky.

About twenty of us stood on the steps of the school, looking out over the city. Not a single word was said. We just stood watching those plumes of smoke rising into the sky. I couldn't believe that the enormous boom we had heard from the fields could have come from the city. It had sounded like it had come from right beside us. I had felt it in the ground. Everything seemed to be silent. Not a bird sang.

Then I heard a distant, strange thud. It was a hard sound to describe. Not a bang. Not a boom. A thud. And there, rising in a thin black line behind the existing plumes of smoke, was another plume of black smoke.

'Jesus Christ Almighty!' I heard a man's voice say, breaking the silence. 'Bombs!'

'Bombs!' The word sank into my head. Bombs. I had seen them on telly going off in Belfast, but here in Dublin?

Everyone stood watching the plumes of smoke rising high into the lovely springtime sky. The man who had broken the silence then looked at his watch.

'Oh, Jesus Christ Almighty ... It's rush hour!' he said. 'God help us all!'

His words rang in my ears. I looked out over the city and watched the smoke still rising and rolling over it. I thought of Da and I hoped he wasn't at a union meeting in Liberty Hall. I thought of Paul; I hoped he wasn't in town buying records. I thought of Clare. She often went 'dossing' with her friends in town on Fridays. All of these thoughts were crashing through my mind. Without even realising it, I was running home.

I got home and found Ma in the kitchen making the dinner. I could hear Clare playing her Rolling Stones records upstairs. As always, Ma had the radio on.

'There's been bombs in town!' I announced.

Ma had her deaf ear to me and didn't hear what I had said. She turned down the volume of the radio.

'What'd you say?' she asked.

'There's been bombs in town,' I repeated.

'Bombs?'

'Yeh, Ma … bombs! I'm telling ya … There's been three bombs in town.'

Ma had been peeling spuds and she dropped the knife into the basin of water.

'The radio said there was some disturbance in town,' she said. 'But they never said anything about bombs. Who told you there were bombs in town?'

I told her about hearing a big one go off from the fields behind St Kevin's and the plumes of smoke I'd seen in the city from the steps of the Sacred Heart.

She stood for a few seconds without saying a thing. The hall door opened and Paul walked into the kitchen.

'I failed me driving test!' he announced. 'What's for dinner?'

'There's been bombs in town,' I said.

'Bombs?'

'Yeh, bombs … and one was really big!'

Ma turned up the radio to hear if anything was being said about it. She looked very worried. Then there came a news bulletin.

'There is an unconfirmed report of an explosion, possibly two, in Dublin city centre,' the man announced.

'I told ya!' I said.

Clare came downstairs and into the kitchen.

'What's for dinner?' she asked.

'There's been bombs in town!' I told her.

'Bombs?'

'Shhh,' Ma said. 'Let me hear the radio.'

We all listened to the radio as Ma put on the dinner. There were constant updates as the news broke. It reported that three bombs had gone off in Dublin city centre.

'Jesus Christ,' Ma said. 'It's rush hour … I hope they gave warnings.'

We continued listening to the radio in the kitchen. We all sensed the seriousness of what we were listening to. Clare and I helped set the table, while Ma checked on the dinner. Da would usually be home by 6 p.m., but by that time there was no sign of him.

Even the people on the radio seemed confused and upset by what had happened. Ma told Clare to keep an eye on the dinner and walked to the front door, opening it to allow a flood of sunlight into the hall. I followed her.

It seemed that everyone else along our road had the same idea. Mrs Redmond was at her door beside us. Mrs Daly and Mrs Reid were also at their doors and, across the road, Mrs Bourne and Mrs Keyes were standing chatting to each other. Everyone seemed to be in shock.

Ma waved across to Mrs Bourne and Mrs Keyes. Mrs Bourne came over to speak to her.

'Is Tommy home?' Ma asked Mrs Bourne. 'And what about Mary?'

'Mary's home okay but no sign of Tommy yet,' she replied. 'And what about Tom?'

'No, no sign of him either,' Ma replied.

Da was a porter in the Mater but his job was driving the ambulance for the Physiotherapy Department. At weekends, he sometimes worked in Casualty when he was doing overtime. If there was ever a crisis or an emergency, all the staff of the Mater Hospital always reported back for duty. When another bomb had gone off in Dublin a couple of years earlier, Da had gone back on duty then. We all knew this and so we weren't worried too much about him. Still, I think Ma would have preferred if she knew for sure. We didn't have a phone and none of our neighbours had a phone either, so Da couldn't even ring them to leave a message. We had been on the waiting list for a phone for about four years.

There was also a bus strike on, so people would be arriving home slower than usual, either on foot or on bikes.

As Ma and Mrs Bourne spoke, I saw Mr Bourne coming up the road on his bike.

'Here's Mr Bourne now!' I announced.

'Oh, thank God,' said Mrs Bourne. She looked very relieved.

'Let me know when Tom gets in, won't you?' she said to Ma as she crossed over the road to meet Mr Bourne.

Ma joined Mrs Redmond, Mrs Daly and Mrs Reid, and they stood with their arms folded talking to each other. They all waved to Mr Bourne as he got off his bike. Nothing was said.

I sat on our front railings and watched. Everyone on the road was either standing at their doors or huddled in small groups talking. All of the Reids and the Dalys were already home. Mr Redmond was off down the country driving a truck so he was definitely okay. The four women stood for another while before Ma broke away saying she'd better get the dinner out or it would be burnt.

It was now well after half-past six and there was still no sign of Da.

'We'll go ahead and eat,' Ma said. 'I'm sure Da has gone back into work. Feck the P&T, it's now I really wish we had a phone!'

She put Da's dinner in the oven and we all tucked in to fish and spuds. I don't think any of us remembered tasting it. We weren't saying a word because we were tuned into the news on the radio. It reported the latest from Dublin and then announced that another bomb had gone off in Monaghan. Reports indicated 'many fatalities'. The word 'fatalities' stirred up all sorts of images in my imagination.

After dinner, Paul, Clare and I did the washing up while Ma went back out to the front garden to watch people coming home. I went outside and sat on the railings again. People were still huddled in little groups or standing by themselves at their doors watching their neighbours coming home. A strange silence hung over the road.

One by one, people came home: mothers, fathers, husbands, wives, sons and daughters. As each person came into view, you could almost feel the relief of these silent watchers. No words were spoken but those who were arriving knew they were being counted in, welcomed home. They'd nod or say hello to the watchers as they passed or went into their homes.

As the evening wore on, the watchers gradually slipped back indoors. By 9 p.m. it seemed that everyone on the road was accounted for. Everyone was home safely.

Ma made a big pot of tea and we helped ourselves to biscuits. We always had pots of tea on the gas in our house and drinking tea always seemed to solve everything. We brought our tea into the front room and turned on the telly to watch the news. In my life, I have never seen Ma pray, but she did say 'Jesus Christ' many times as the TV revealed the horrors of what had happened. They were shocking images.

Talbot Street was where the big bomb had gone off, right outside Michael Guiney's shop. This was where Ma bought towels, curtains and everything else you'd need for a home. It was my worst nightmare as a kid. I remember standing in the shop as Ma had curtains measured out or checked out the patterns of wallpaper (which seemed to be one of her hobbies). Then there was the famous routine of 'bringing things back'. Ma was a master of bringing things back. I would stand beside her listening to her demonstrating how an item was not exactly what she expected or how it did not do what it was meant to do. She could convince anyone towards her point of view.

Da's friend Joe McBrien worked in Guiney's and Ma was worried that he might have been hurt. The news reports said that many people had been killed. It was hard to believe.

I closed my eyes and thought about all of those people who had died at the very moment that the 6th-year pupil was about to score a goal against the teachers. I thought of that boom, that enormous boom that scared the birds from the trees, silenced the singing blackbird and silenced a whole field of screaming fellas watching a football game. I remembered witnessing that third bomb go off from the steps of the Sacred Heart – that dull thud and the plume of smoke. This was another moment when people had died. My mind found it all too difficult to take in.

After the news, Mrs Bourne called over to see if Da was home yet. Ma told her not to worry, as he had no doubt returned to the Mater to help out.

We watched telly. We had more tea. Most of all, we waited for Da to come home.

It was almost midnight when we heard the key turning in the hall door and Da walked in. He looked very tired and pale.

'Is everyone okay?' he asked Ma.

'Yes, everyone around here is safe and sound,' she replied.

The poor man had barely time to sit down before I began bombarding him with questions. Ma went out to the kitchen to warm up his dinner, which he ate very quickly. He was starving. She told us all to let Da have his dinner in peace.

After his dinner, another pot of tea was made. I sat down across from Da. I sipped my tea and allowed him drink some of his.

'Where were ya, Da? Were ya in the Casualty?'

'I heard the big bomb go off, Da,' I continued before he could answer my question.

'And I saw the third bomb go off from the steps of the Sacred Heart,' I added.

'I heard the big one go off too,' he said. 'I was in Phibsborough and I felt the ground shake under my feet.'

'And did you know it was a bomb?' my interrogation continued.

'No, not really,' he replied. 'But then I heard all of the ambulances and fire brigades and I knew something big had happened. So I rushed back to the hospital to see if there was anything I could do to help.'

'So, were you in Casualty?' I asked again.

He shook his head. 'No, they asked me to drive doctors and nurses down to assist with the injured in Talbot Street,' he said quietly.

I was shocked. 'When were you there?' I asked.

Da sighed. 'I was there before half six,' he said.

I had seen pictures of Talbot Street on the news. It looked horrific. 'Was it bad, Da?' I asked.

'Bad?' he replied.

He shook his head.

'I hope I live long enough to be able to forget what I have seen tonight.'

He sighed again.

'Let me tell you, no cause is worth the carnage I have seen in Talbot Street this evening … No cause!'

His words had a huge impact on my young mind.

I couldn't think of anything to say, except 'Will you have another cup of tea?'

He smiled as he accepted my offer and I poured him a fresh cup of tea.

'That's a darling cup of tea,' he said as he sipped the hot brew.

Chapter 7

THE WINDHOVER

'**W**atch the Duck … he's a sneaky bollix,' Johnny had warned me before the summer.

Johnny's words echoed in my brain as, in September 1974, I walked around the corner to commence my secondary education in St Kevin's, which was run with a firm hand (and fist) by the Christian Brothers, with the Duck as head brother.

I had done the entrance exam the previous May and apparently passed with flying colours. I was informed that I was to be assigned to the A class. The classes were graded and you were assigned to a class according to how clever they thought you were. The A class were the swots, the B class were the middle-of-the-road fellas and the C class were not considered the most studious but they learned trades like woodwork. This system put everyone in his place from the very beginning.

I was no swot and the thoughts of sitting among a load of swots did not appeal to me. As well as that, I wanted to learn German instead of French, which you couldn't study if you were in the A class, and that provided me with an excuse to be transferred into the B class.

All of us first-year students started a day earlier than the other years. We all gathered in the yard at half nine in the morning and were met by the Duck. He was a small, stocky, bald man. He wore a black jacket, trousers and shirt. His full white collar always seemed too tight around his neck and his neck bulged slightly over it. He glared at all of us through his black-rimmed glasses. He had steely blue eyes that would cut you in two.

He called out each fella's name and ordered him to stand in one line or the other. When we had finished, he pointed to each line and said, '1A, you lot; 1B, you lot; 1C, you lot.' With that, he marched each line into its respective classroom and told us to find a seat. We took our seats in the cramped classroom. I was lucky to get a seat by the window half way up the room. The bottom half of the window was frosted like the windows you'd see in a toilet. It meant you couldn't see out. He informed us that this was to be our classroom for the year, that we were to remain in these seats at all times unless told to move by a teacher, and that we were to keep the classroom neat and tidy.

He then stood at the top of the class and made a short welcoming speech. 'You're here to learn,' he said. 'And learn you will! I won't take nonsense from any of you. None, do you hear me?'

No one said a word.

'Do you hear me?' he growled.

'Yes, brother,' we answered in unison.

With that short, warm welcome, he was gone to terrorise 1C.

Class 1B contained lots of new potential friends. There were also lots of fellas from my class in primary school so it was an easy transition. It did take a while to get used to the new system where we had different teachers coming to us for each subject, or us going to the art room or science lab. We also had new subjects like science, German, art and business studies. In all, we had nine subjects as well as civics, which was meant to teach us about politics and our civic duty to our country. The civics book was opened a few times during the first week but never again. Our civic duty was not as important as playing GAA football, in the eyes of the Christian Brothers.

Over the first week, we were introduced to our teachers, some of whom were Christian Brothers. These included some interesting characters such as 'Dunderhead' (named after his habit of belting you on the crown of your head with his knuckles if you got a question wrong in his class and announcing that you were a dunderhead as he did it), 'Cat Weasel' (named after the TV show at the time), 'Juno', 'Spike', 'Dracula' (he wore a teacher's gown), 'Smitty', 'the Silver Fox' and, last but by no means least, 'Black Benny'.

Black Benny was a Christian Brother. He had thick black hair that was going grey at the sides. It was greased back with oil. He wore dark glasses and you could never see his eyes. He didn't teach us, but he was in charge of our class. If we had a sick note, we had to give it him. Black Benny was always the one who did the supervision in the jacks during our morning breaks. He was always the one who supervised the fellas when they were changing for PE or for football. If he chatted with you, this would inevitably lead to a slagging later on.

'Jaysus, Dempsey, ya must have had a bath last night. Black Benny likes ya big time this morning!'

It was all good fun and part of being in the school.

My first real experience of Black Benny was during the first week of my first year. I was late back from lunch and was running across the yard. I saw him crossing the yard as I ran. As I passed him, he stuck out his arm and thumped me in the chest. This knocked me off my feet. I picked myself up and stood there with my mouth open.

'No running in the yard after break,' he snarled.

I looked at him. 'Are you a Christian Brother?' I demanded. I said it with real aggression.

Bang! I got another thump in the chest. I was grounded again. I got up.

'Are you a Christian Brother?' I again demanded.

Wallop! He hit me across the face. I half blocked his slap. It hurt but I didn't fall.

'Don't use that tone of voice with me, ya young bucko,' he shouted angrily. 'What's your name?'

'Eric Dempsey is my name and I just want to know if you're a Christian Brother,' I replied with real indignation. 'Because if you are, then I must say that wasn't a very Christian thing to do!'

He was gobsmacked by my cheek and launched another blow, which I managed to block fully.

'I don't think Jesus would be very happy with one of his Christian Brothers doing that to a young fella,' I said, looking straight into those dark glasses of his.

I stood my ground. Years of debating stuff at home had taught me well. He didn't know what to say. He grunted and shoved past me. When he had gone, I almost collapsed in a heap from the number of belts he had given me but also from shock at my own cheek.

That evening I spoke to Da about the thumping I had. Da was furious but was very proud of how I had stood up to Black Benny.

'The pen is mightier than the sword,' he said. 'This Brother Black Benny has learned a hard lesson. He's discovered that when it comes to us Dempseys, it's the tongue that's mightier than the sword.'

He felt I had defended myself well and had challenged him with logic. He reckoned that Black Benny probably wouldn't cross me again. He was right: Black Benny never messed with me after that. There was no need for Da to do anything.

That wasn't the case with another teacher I encountered who was football mad. In our school, they only really cared about you if you could kick a football. Many past pupils of the school went on to win All-Ireland medals with the Dublin football team. If you were the next Albert Einstein or Isaac Newton, or had the talent of Beethoven or van Gogh, the Christian Brothers weren't interested in you. If you were Einstein, Newton, Beethoven or van Gogh and could kick a football, I have no doubt that your other talents would never have been discovered. As for me, well, sports weren't my big thing. So I was not that popular with the GAA selectors and trainers.

This particular teacher's distaste for me came to a head one day when we were out on the football pitches. I was standing in line waiting my turn to do something or other – probably run and catch a ball that was kicked to me. However, I became distracted by the sight of a hunting Kestrel at the bottom of the field. It hung into the wind and remained motionless. It fanned and un-fanned its tail as it adjusted to the slight variations in the wind direction and speed. It was a male, with a russet-red back and a grey and black tail. It went to dive but stopped in mid-air. It hovered again. Unlike me, it was the picture of concentration. It adjusted its position once again and swooped down into the tall grass along the bottom of the field. It disappeared for a few moments before rising again with a small mouse in its talons. It was the first time I had ever seen a Kestrel catch its prey. I was transfixed.

The problem was that I was so transfixed, I didn't realise that I was being shouted at. I was delaying everything. I only became aware of the teacher when he was almost beside me. He had run about a hundred yards to reach me.

'Dempsey, are you fuckin' deaf?' he shouted. 'What are ya doing? Get out of that line and come over here.'

I walked over to face him. His face was red with anger and his eyes were bulging.

'Are you deaf? Did you not hear me shouting at you to take the ball?'

'No, sir,' I answered. 'I was looking at a Kestrel hunting.'

'You were looking at birds?' he snorted, laughing out loud. 'Are you a feckin' pansy or wha'?'

'No, sir, I'm an ornithologist,' I answered confidently.

With that, he gave me a punch straight into my stomach. I fell to the ground, winded, gasping for breath. He reached down and grabbed me by the hair and lifted me up. I was bent double. He held me in that position and kicked me three times in the face. He knew how to kick hard. He let go of my hair and I fell to the ground again. I got one more kick into my chest.

'Now get up and get to the back of the line. I'll teach you to be watchin' feckin' birds. I'll make a footballer of you yet!'

I slowly got up but, instead of going back into the line, I walked past the line, out of the field and back to my class. I heard him shouting at me to come back as I left. I ignored him. I picked up my schoolbag and, still in my football gear, I walked out of school and went home.

Ma worked for a few hours each day and came home late in the afternoon. She was surprised to see me. When I told her what had happened, she was furious. We agreed that we would wait until Da came home and then we would hold a family meeting. After tea that evening, I recounted to Da, Paul and Clare what had happened. I didn't exaggerate. I didn't have to. My face was bruised and I had a big purple mark on my chest from his final kick. Paul was enraged. Clare said that we should all go around and tell him what we thought of him. Da said he would look after it.

I returned to school the following morning. The Duck saw me and called me over.

'I hear you were causing trouble yesterday,' he said. He was looking away from me as he spoke. 'And then you went mitching in the afternoon.'

'If you want to know about what happened, you can ask my father who is coming around to discuss the matter with you this morning,' I said.

This is what Da had told me to say. The rehearsed statement threw the Duck off guard. He walked away without another word.

That morning, Da turned up at school when we were on first break. We were all in the yard and everyone in my class was delighted to see Da walk in the school gate. Fellas greeted him as if they knew him.

I watched from a distance as Da first took the Duck out of the staff room. I could see he was giving him a tongue lashing of the sort that only Da could give. The Duck was standing without saying a word. Next, 'himself' came out of the staff room. He strutted out with his cocky head on him. He was a tall man, but within seconds I could see him visibly shrink before Da. The whole encounter lasted about five minutes.

I then saw him offer Da his hand. Da refused it. He walked out of the school and went home.

After break was over, I went back to my class but was very soon after summoned to the Duck's office. I walked in to find this teacher standing there. He glared at me as I walked in.

'You have something to say to Mr Dempsey, don't you?' the Duck said.

'Yes,' the teacher said quietly. 'I am sorry for hitting you yesterday.'

He offered his hand. I shook it.

'I offered my hand to your father,' he said, 'but he refused it and told me it was you I should be offering my hand to, not him.'

I smiled to myself as I imagined Da saying those words.

When Da got home later that evening, I asked him what he had said to the teacher. He simply said, 'I think he might have discovered that when it comes to us Dempseys, the tongue is mightier than the sword!'

Then he asked me about the Kestrel I had seen. With the distraction of the bashing I'd taken, he had forgotten to ask me about it. I told him all I knew about Kestrels, the differences between males and females, and how they hovered when they hunted.

He smiled. 'Ah, the Windhover,' he said.

He then recited the first verse of one of his favourite poems, 'The Windhover', by Gerard Manley Hopkins.

I caught this morning morning's minion, king-
dom of daylight's dauphin, dapple-dawn-drawn Falcon,
    in his riding
Of the rolling level underneath him steady air, and striding
High there, how he rung upon the rein of a wimpling wing
In his ecstasy! then off, off forth on swing,

As a skate's heel sweeps smooth on a bow-bend: the hurl
 and gliding
Rebuffed the big wind. My heart in hiding
Stirred for a bird, — the achieve of, the mastery of the thing.

I listened carefully to the beauty of the words and the rhythm of the verse. There was no doubting that the pen was indeed mightier than the sword.

Driving along the motorways now, I often see Kestrels hovering over the grassy edges and my mind occasionally drifts back to the day when I got that bashing. However, it's not an incident I remember in any negative sense but more with a deep sense of pride.

When I needed it most, Da had stood up for me. It was more than just the fact that he would not allow any teacher to assault me like that. No, it was much more than that. Da would not allow anyone to put me down. Da was proud of the fact that I took such an interest in the world around me. He was proud of who I was and of the person I was becoming.

Chapter 8

'IT WAS THE RAINBOW GAVE THEE BIRTH'

Living in Finglas, we really were lucky to have the Botanic Gardens ('the Bots') very close to us. When we were kids and teenagers, the Bots was our playground. It is now time for a confession: the Bots was also the ideal place for mitching from school. The thoughts of facing an afternoon of the Silver Fox for religious studies or Dracula for another go at my 'trial balance' was always enough of an incentive to mitch.

Clare was a great accomplice in my mitching as she had developed a very special talent: she was a master forger of Da's beautiful handwriting. I handed many of her notes up to Black Benny that said I was attending a doctor's appointment and so couldn't attend classes in the afternoon. Comparing the genuine article to her forgery, it was impossible to tell them apart.

So, with such talent to rely on, that aspect of skipping school was always well covered. There was little risk of being seen by any of the neighbours in the Bots, and if any of the gardeners or security men challenged me as to why I wasn't at school, I always had my well-rehearsed answer ready and I delivered it with great sincerity:

'I'm doin' a biology project.'
It worked every time.

Looking back through my old notebooks, I see that I document
the first date on which I saw a Kingfisher with binoculars and a
telescope as 20 August 1978, at Broad Lough, near the town of
Wicklow. However, my very first encounter was in fact several
years earlier, and in most unusual circumstances. This encounter
with a Kingfisher was during one such occasion when I was 'doin'
a project' in the Bots.

No matter how many times I've seen a Kingfisher, when I
come across one it always stops me in my tracks. Even if I know
that the birds are in the area, seeing one always gives me a sense
of the unexpected. It is like I am seeing the bird for the very first
time. Most times, the only view I might get is a flash of turquoise
shooting upstream. Occasionally, one might perch long enough
for me to enjoy the beauty of its bright orange underparts and
royal blue wings and crown. When the bird is in flight, it is that
bright turquoise back and rump that catches the eye. Unlike many
species, both males and females are identical in plumage; the only
difference is that males show an all-black bill, while females show
a reddish base to the lower mandible (the lower part of the bill).

Common Kingfisher (their correct name) is a widespread
species in Ireland, found along streams and rivers, lakes and ponds,
and even on coastal estuaries. They are not common but they are
not rare either. The species is found all over Europe, the Middle
East and Asia, hence the name 'Common' Kingfisher.

On this particular winter's day in the Bots, it was freezing and
the ground was covered by a good dusting of snow. The ponds
at the back of the gardens were frozen over and the snow was
undisturbed except for bird tracks. I was mitching off school in
the hope of seeing birds. Cold weather might have brought in
winter thrushes or, even better, a Waxwing. I did see Redwings and
Fieldfares as well as the other usual garden birds.

It wasn't long before the freezing weather forced me to take refuge in one of the 'hothouses', as we always called the large glasshouses. It was warm and humid inside, a complete contrast to the biting cold outside. In one of the larger glasshouses, there was a large raised pond surrounded by exotic plants, with water lily pads floating on the surface. The pond itself housed a good collection of goldfish, which added darts of colour to the dark green of the water.

On this cold day, however, I was in for a very special treat. As I approached the pond, I saw something bright and turquoise flash up from the edge of the pond and into the canopy of plants above.

I stepped back and waited. My heart was thumping. Following what seemed like an hour, the flash of turquoise appeared again, splashing into the pond only to emerge from the water in the form of a Kingfisher. It had a goldfish in its beak. I was spellbound. It flew off up the corridor and into the next glasshouse, which was higher and larger, and disappeared. I followed it and found it sitting high in the canopy of the tall trees. I felt like I was in the rainforests of Papua New Guinea watching a bird-of-paradise. I couldn't believe my luck. It was a bird experience like no other.

This bird had somehow managed to find its way into the glasshouses during the coldest snap of the year. Inside, it basked in a warm, humid, tropical rainforest climate while the birds outside froze. Here, it had enough goldfish to keep it going for weeks.

Years later, I spoke with one of the gardeners about this bird and he told me that it had been inside the glasshouses for about three weeks until they managed to catch it and release it along the river, but not before it had eaten every single goldfish in the pond.

I remember telling Da about it later that evening. I was so excited about this bird experience that I forgot he might ask why I was down in the Bots and not at school. He never asked that question. Instead, he turned to his vast store of poems in his head to find one that fitted the moment perfectly. He thought about it for a few seconds before reciting 'The Kingfisher', by the 'nature poet' W. H. Davies, which begins with the lines:

It was the rainbow gave thee birth
And left thee all her lovely hues.
And as her mother's name was tears
So runs it in thy blood to choose
For haunts the lonely pool, and keep
In company with trees that weep.

Whenever I see a Kingfisher, I am reminded of my school days and that magical moment in the glasshouses of the Bots. I am also reminded of sitting in the kitchen with Da as he quoted 'The Kingfisher'.

Perhaps ultimately seeing a Kingfisher reminds me that, as Davies himself wrote, 'I also love a quiet place that's green, away from all mankind.'

Chapter 9

FLEDGING

O f course, the best thing about secondary school was the long holidays, at Christmas, Easter and especially during the summer when we had three months off. These were long months of bliss: no homework, no schoolbooks and no sitting in classrooms. Instead, we had freedom to roam.

Da always told a joke coming up to summer. It still makes me laugh to this day:

'Where are you going for your summer holidays?'

'I'm going to Rome!'

'Rome?'

'Yes, I'm going to roam around Finglas for the summer.'

That was exactly what I did during the summer: I went roaming around Finglas. It was a time for looking for birds and a time for hanging out with friends (as well as getting to know some of the young wans of the area). Of course, there was also the Bots and Glasnevin Cemetery to wander around. The old part of the cemetery was a wonderful place for nature where history mingled with the wilderness.

The dread of the end of the summer holidays is something everyone knows only too well. It's a countdown of weeks, then days, until that sickening day arrives: the very last day of the school holidays. You do your very best to make the most of that last day of freedom before you face into another school term, more homework and hours of sitting in the classroom. And somehow, within a few days of returning to school, it seems like you never had that summer of fun.

However, despite the occasional thump from a teacher, I found secondary school grand in most regards. There was a sound gang of lads in my class and we had great fun in between learning. We also had great fun whilst learning too.

I can imagine it must have been a tough school for teachers. I remember witnessing a maths teacher punching one of the bigger fellas in the class only to be punched back even harder. We had to help him up off the floor. We had teachers who were stricter than others and they seemed to keep control. 'Softer' teachers provided fun times for the lads and often the class would disintegrate into chaos. I vividly remember a 'stand-in' English teacher, who was obviously fresh out of college (he was not much older than us), having his *Exploring English* textbook taken from his hands and flung out of the classroom window. The poor teacher didn't know what to do. He walked out of the classroom, retrieved his book from the grass beneath the window and never darkened the door of our classroom again. I wonder if he gave up teaching after he encountered our class.

For some reason, the academic side of secondary school was an easy enough challenge for me. I had an interest in many subjects, including English and history. I could somehow do maths without much pain, always managed to get my trial balance to work out in accountancy, liked geography and science, and even took to Irish, due to a brilliant teacher called Spike who had us all loving and speaking the language. I was even lucky enough to have a good grasp of German and found that I could paint and draw well enough to get by in art class.

I even had the pleasure of winning many poetry-writing competitions in English class. Mr McCarthy ran these fortnightly competitions and I think he was more than surprised to discover that a dosser like me could write poetry. My great-great-grandfather, James Sutton Jackson, was a poet who published a book of poetry in 1843. My Uncle Gerald was a poet. Ma was also a talented writer of letters and poetry. So perhaps some of these combined genes had something to do with me being able to write a bit of poetry. I loved winning these competitions because the prize was that I'd be let off English homework for the weekend.

The truth was that I never studied. I reckoned that if I couldn't take in a lesson during the day then cramming it all in when I got home would not do anything for me. That was my logic, but of course I spent my time reading other books and consuming bird and wildlife magazines. That was the only study I considered truly worthwhile.

I was lucky in that at home we engaged in long and heated debates and discussions on everything from the unravelling situation in Northern Ireland to the existence of God, from Russia and Communism to the various global religions, and from the start of the universe to whether there was life after death, and every other subject in between. These discussions would start after or during dinner, which was always around teatime. Something we'd heard on the news might spark the discussion and in minutes we'd all start giving our tuppence-ha'penny worth. These long debates sometimes continued on for hours and suddenly we'd find it was 10 p.m. We might have had lots of homework to do, but Da would write us a letter to have us excused from having to do it.

'You've learned more from giving your views and debating these ideas than you would have from doing homework,' he'd say.

He was right. Such long debates and heated discussions helped form our early views on life, informed us of the world around us, got us thinking for ourselves and helped us to stand up for our own beliefs. It was an education that no school could ever give.

There was never any pressure on any of us at home regarding exams. We were always told to do our very best. I always tried my very best when doing exams and, somehow, this tactic always worked for me.

In June 1977, I did my Intermediate Certificate exam. While other fellas were bundles of nerves, I was very relaxed. This relaxed state was not from a feeling of arrogance or knowing it all but from the fact that I felt no pressure from Ma or Da other than to do my best. I did my nine papers and headed off into the summer months without thinking about it again.

I returned to St Kevin's in September 1977 to start 5th year. My Inter Cert results came out and I got eight Cs and one B (in maths of all things). I was surprised at how well I'd done. Even the Duck muttered something that sounded like 'good results' to me as he handed me my sheet of results. Dracula, who taught me accountancy, took me aside and lambasted me:

'Look at those results,' he said. 'If you'd bother your arse studying, imagine what results you'd get!'

I couldn't argue with him; no doubt he was right. I can imagine he felt frustration at what he perceived as laziness on my part. Of course, what he didn't know was that I was studying my brains out each night, but it was bird and wildlife books I was studying not school textbooks.

With the Leaving Certificate now our main aim, we had to choose our subjects. This was when, in the first days of 5th year, school began to disintegrate into a situation which would ultimately bring me on a whole new path in life.

Irish, English and maths were compulsory subjects. I was delighted when I was put into Spike's Irish class. However, the first part of my disillusionment with secondary school started here. Spike had too many students in his class so it was decided that ten students should be selected to be removed. This was done randomly: the first ten in the front row of his class were taken out. I was amongst them. I was gutted to be leaving Spike's class, but I was even more disgusted to learn that I would now have to endure

Black Benny for two years of Irish. Within weeks, I'd lost my love for the Irish language.

I thought it could not get any worse, but it did. I loved history and had chosen it as a subject for my Leaving Cert. I looked forward to learning more about what had shaped Ireland and the world. However, too many pupils had opted for history. In order to resolve this dilemma, all the names were put into a hat and those whose names were pulled out were taken out of the history class and put studying geography. My name was pulled out and I ended up in the geography class.

Finally, just as I thought this run of bad luck was over, it emerged that there were too few students who wanted to do accountancy. As someone who could do a trial balance with my eyes shut, I had opted for this subject. Alas, all of us who were studying accountancy were lumped in with the business studies class. This subject wasn't too bad, but the teacher was soft and his class was just one big jumble of fellas talking and messing. Nothing would ever be learned.

By the time Christmas came, I had lost all interest in school, a fact that was reflected by my end-of-year exam results.

When the new year commenced I really did try my best to be enthusiastic about my education, but it was just not for me. I starting spending more time in the Bots 'doing biology projects', and Clare's wonderful forgeries of Da's handwriting were in greater demand than ever. The school's career guidance teacher, a lovely, gentle little nun, was convinced that I should go on to third level and study zoology. She said that if I concentrated on my studies, this could be my path in life. I even liked the idea but found it hard to imagine enduring another eighteen months of secondary school.

Then, in February, during the mid-term break, salvation arrived at my door. It was a brown envelope with a harp on it. It was from the Civil Service Commission. In 3rd year, we had all been encouraged to do the civil service exams. I remember sitting one for Junior Postman. I had done it just to get the day off school. I opened the letter and read with great interest. I had been successful in the exam for the position of Junior Postman and had been placed number 23 on a panel of candidates. It went on to say that I would be called to take up the position in the coming months. I almost cheered out loud. I had a way out of school. I tucked the envelope into my pocket.

That evening after dinner, I showed the letter to Ma and Da. They were very pleased that I had done so well in this exam but were none too pleased when I suggested that I should leave school and take up this job. I argued the pros and cons with them. A big plus was that it was a civil service job, a secure, permanent and pensionable position. But they were obviously disappointed that my studying something like zoology might be jeopardised by my taking a job that wasn't really what I should be doing. They were disappointed that I wasn't even considering doing my Leaving Certificate. I understood their concerns so I suggested a perfect compromise.

'I'll view this job as a summer job,' I offered. 'Then, when September comes around, I'll start 5th year again and hopefully I'll get some of the subjects I want to study and go on to do my Leaving Cert.'

It was a very sincere offer on my part and one that was accepted by my understanding and supportive parents.

'If you want to give up school and start to work, we'll support you fully,' Ma said. 'You're old enough to make your own decisions.'

I expected a 'but' at the end of her sentence, but there was none. They both trusted my decision and empowered me to make up my own mind.

It was done. I rang the Civil Service Commission and confirmed that I would be taking up their offer of the position of Junior Postman (JP). When the mid-term break ended, I did not go back to St Kevin's. I took time off school and enjoyed a good break.

The day I first entered the workforce was 10 April 1978, and it was a cold windy day with sleety rain falling in horizontal sheets. I was 16 years old. Once my details were processed in the offices of the Civil Service Commission, I was assigned to the Central

Telegraphs Office (CTO) in the General Post Office (GPO) in Dublin city centre.

My first job was delivering telegrams around Dublin on a bike. Yes, you read that correctly … telegrams. I was given a postman's uniform but my jacket had special hooks around the waist that allowed for a thick leather belt on which hung a leather pouch where the telegrams could be stored safe from the wind and rain. Before being allowed to deliver any telegrams, I had to do a 'bicycle test' to make sure that I could ride a bike safely. The boss man, an aul' fella called Fergal, stood watching me riding a big old jalopy of a postman's bike up and down Prince's Street. Once he saw I was able to ride a bike, I was sent for two weeks' training with another JP to learn the main streets and government department buildings. It was great fun.

I earned £26.77 that first week and it was with a great sense of pride that I handed up half of my wages to Ma. I was a working man now, and if I was staying at home I would pay my way.

Within two months of starting in the CTO, I was transferred to an indoor position. I was still based in the GPO but was now working for Paddy Cahill on the internal post. Basically my job was delivering internal mail to all the offices around the GPO. It was such a cushy job. Even better was the fact that my arrival into the world of the GPO coincided with the appointment of the first Junior Postwomen. It was like going to a mixed school, except with no teachers or homework and I was earning money. I was in heaven.

Within the first month, I had bought myself a new pair of binoculars: Russian Helios 7x50s. They weighed a tonne but they were my own decent pair of binoculars. By early summer I had saved enough to buy a Kowa telescope. I ordered it from a shop in London and Paul, who was on holidays over there, picked it up for me. It was like Christmas when he arrived home and I took out my scope for the first time. I was spending money as fast as I earned it. Now I had superb optics to go along with my books and my subscriptions to journals. Even better, I had no homework to worry about, was dating a lovely girl from the typing pool in the GPO and was having great fun.

Sadly, my first real girlfriend broke it off with me when I chose a weekend's birdwatching on the Saltee Islands in Co. Wexford over spending a weekend with her.

She deserves better than me, I thought to myself as I enjoyed watching my first Puffins.

The months passed, and as the last week of August approached I knew I had to do some thinking. I had made a bargain that I would consider this job a summer job and now it was crunch time. But here I was with money, a fun job with a load of new friends, no Black Benny or the Duck to worry about and no homework. By the time the conversation with my parents took place, I had already made up my mind: there was no way I was going back to St Kevin's.

I think both Ma and Da suspected that this would be the case but argued strongly for how important it was for me to get a good Leaving Cert. They were right. A good Leaving Cert in 1978 was like an honours degree now. It offered superb opportunities for the future. So I struck a new accord with them: I would study for my Leaving Cert at night. This was even encouraged by the civil service, which would subsidise my evening courses and give me study leave and time off to do exams. It was the best of both worlds: I still had my well-paid job and would study for my Leaving Cert. So, in September 1978, I began studying for my Leaving Cert at night. I almost enjoyed the evening classes.

At the end of the first week of evening classes, I was sitting at home with my study books and homework. I had English and Irish essays to do. I had a heap of maths questions to complete. I had to write something about the War of Independence, had a trial balance to complete and had to read something about plants. I looked at all of this material that required my absolute attention. It was September. The autumn migration was starting. It was the weekend. My scope was on its tripod in my room. It seemed to be calling to me to bring it out into the wilderness of the North Bull

Island. Instead of doing my homework, I started reading my bird books and dreaming of what birds I might see if I went out.

I looked at my homework and I looked at my scope. My scope won.

I never did my Leaving Cert.

A little over fifteen years later, a very special day arrived. I went up home to have dinner with Ma and Da. After dinner, I asked if I could see them in the front room. They were mystified. I suspect they thought I had something bad to tell them.

They followed me in and I asked them to sit down. I took a large, padded envelope from beside the sofa where I had put it earlier. They had no idea what it was.

I smiled at them.

'Here's me Leaving Cert!' I said, as I handed them both the envelope.

They opened it together and pulled out the advance copy of my first book, *The Complete Guide to Ireland's Birds*.

'It's the very first copy off the press,' I explained. 'I want you to have it.'

Ma and Da were chuffed. They hugged me. Da opened the book and browsed through the pages before he smelt it (he liked the smell of new books almost as much as he liked the smell of old ones).

'Thanks for believing in me,' I said to them.

'We've always believed in you,' they replied in unison.

In my heart, I knew they spoke the truth.

Chapter 10

ON STRIKE — THE MAKINGS
OF A BIRDER

By early 1979, I was birdwatching every weekend and making copious notes about the birds I was seeing. I was just seventeen years old and, having bought my brand-new angled Kowa telescope the previous summer, the whole world of birds had opened up to me. This telescope was so much easier to look through than the 'straight through' draw-tube telescopes that many birdwatchers still used. Now, distant blobs that had been feeding out on the mudflats on the North Bull Island were revealing themselves as Black-tailed Godwits, Redshanks and Golden Plovers. I had joined the Irish Wildbird Conservancy (IWC) (the old name for Birdwatch Ireland (BWI)) and had gone on my first field trip with them to the Blessington Lakes in Co. Wicklow on 14 January. It was wonderful to be in the company of a few birdwatchers of my own age as well as the more experienced birdwatchers who were keen to share their knowledge.

In January, I had also started attending a bird course run by Vincent Sheridan in the college in Marino. At the time, he was the chairman of the IWC and his course took us through many fascinating aspects of birdlife. I used to look forward to Wednesday evenings and cycling down to Marino to spend two hours in class learning about birds. To my mind, this was better than doing the Leaving Cert any day.

It was here that I first met Don Conroy who, at the time, was still working in an advertising firm and was also doing a lot of stage acting. He was (and of course still is) a very talented artist and we struck up a close friendship that lasts to this day. I was also sketching and drawing the birds I was seeing. Drawing birds really helps to capture their details, especially if you attempt to draw them 'in the field'. My old notebooks are filled with a selection of drawings ranging from dire to not so bad. Don was very encouraging about my artistic endeavours.

In February of that year, I also met Killian Mullarney at a bus stop in Malahide. Killian was the rising star of birdwatching in Ireland. He had seen two Slavonian Grebes that day off Portmarnock, birds I had never seen. I sat with him on the bus journey into town listening to him recount the finding of Ireland's first Ring-billed Gulls in Belmullet in Co. Mayo. He pulled out his notebook. It was a black Daler hard-covered sketch pad and it revealed immaculate pencil drawings and watercolour paintings of these gulls which, until then, I had never even heard of. We also discussed the shocking news of the shooting of Ireland's first Belted Kingfisher in Ballina. Killian had actually seen this bird only the day before he had found the Ring-billed Gulls. The poor kingfisher had been first seen in Ballina in December. Killian and a few other lads travelled to see it on 3 February and had found it along the river. That afternoon, a local taxidermist shot it. The shooting of this bird made the news. He then showed me sketches and paintings of the kingfisher. They were stunning. I didn't show him my own terrible little pencil sketches in my notebook. However, for an inexperienced birdwatcher like me, it was great to be sitting talking 'bird' with one of Ireland's leading bird experts.

So, in early 1979, life was good. Birdwatching was good. I was enjoying working, having my own money and keeping company with a lot of new friends, both in birdwatching circles and at work.

Having been raised in a good trade union house where the great labour struggles, the 1913 Lockout and Jim Larkin were all part of our upbringing, I was of course a member of the Post Office Workers' Union (POWU). Within the Department of Posts and Telegraphs, there were a number of unions for the different grades. The clerical grades had their own union, while technicians and engineers were in another. Paul, who was a technician in the Department of Posts and Telegraphs, belonged to the latter trade union. The POWU covered the postal workers like Junior Postmen (and women), Postmen, Postal Clerks and Sorters. I joined the union, as did everyone else, on the first day I started work in 1978. Little did I realise it, but the POWU would have a very important role in my transition from a novice birdwatcher to a known young birder in my own right.

During the early weeks of 1979, there were rumblings of discontent among the postmen and clerks. Pay increases and changes in working conditions and shifts were the most frequent subjects of discussion at tea break among the more senior men. For us JPs, these things seemed irrelevant. True, when we turned eighteen, we would be sent as rookie postmen to the dreaded dark, dirty and run-down dungeon that was the main postal sorting office in Sheriff Street. For the moment, Sheriff Street was to be avoided at all costs, since it meant being the lowest of the low in seniority among the postmen and being given the worst shifts, which usually entailed starting times of half five in the morning, working weekends or doing night shifts.

In the second week of February, POWU meetings took place at which some strike action was planned to highlight the demands of the workers. At one such meeting that I attended, I learned that it had been decided that the union would hit the Department during one of the busiest postal times of the year outside of Christmas: the week of St Valentine's Day. We were told that this would be an official strike and that we would all lose one day's pay as a result. Pickets would be placed on Sheriff Street but JPs were told that we were not required for picket duty. For me, it meant that I would have a day off work. I felt how a school kid might have felt if they turned up for school only to be informed that the heating was not working and school was cancelled.

At home that evening, I sat and spoke with Da about the impending strike. Da was the shop steward of his union in the Mater and he had great knowledge of all things to do with trade unions and strikes. In fact, because he was such a skilled trade union negotiator, he was offered promotion into the management levels of the Mater where he would be required to represent the management's side to such negotiations. This would have offered him better wages and conditions but Da turned it down. He felt it would have been a betrayal of his fellow workers.

'The only things that should never be bought or sold are your principles and your integrity,' he said. 'These are what define you and they should never be compromised.'

So Da fully supported the strike action. The workers in the postal services had very poor working conditions, had little or no shift pay and did not receive good wages. In Da's opinion, the unions, having tried to negotiate for changes, had no choice but to strike.

The day of the strike was a cold stormy one. I spent the day around Dún Laoghaire in south Co. Dublin. Inside the harbour, I had the best views I ever had of Red-breasted Mergansers and Great Crested Grebes. Birds seen on an unexpected day off always seemed so much better. It was a long and a good day.

The following morning, I returned to work. Apparently the unions and management were meeting to discuss the workers' concerns. Over the following week, things seemed to go back to normal a little, but by the end of the third week of February negotiations had broken down. Another one-day all-out strike was called. So, for the second time that month, I had a day off.

This time, however, I was on picket duty. It was a strange experience to walk up and down Prince's Street beside the GPO carrying a placard. The older men were impressed by us few JPs. We were suddenly 'comrades'. One or two of them even called us 'brother'. I felt transported back to the time of the Lockout and

Jim Larkin. I was a comrade in the Russian Revolution. People stopped in the street and spoke to us, encouraging us and wishing us luck. Cars and motorbikes that passed us coming from the *Evening Herald* depot beeped their horns in support. We cheered and waved each time they did this. It was a great experience for a young fella. The down side of the day was seeing other Department workers from different trade unions casually walk past our pickets. It was the first time that I truly understood what Da had spoken of when he said that, in our lives, none of us should ever pass a picket. Paul worked in another depot and so this didn't yet pose a challenge for him.

The day passed and I again returned to work the following morning but this time word came that we were not returning. It had been decided that the strike would last until the demands of the union were met. It was 'all-out' indefinitely. I was assigned picket duty in Prince's Street beside the GPO from ten in the morning until two in the afternoon every Monday. After that, the time was my own. No one knew how long this would last.

For me, this was the best news ever. Of course, I was young. I had no responsibilities in the world. For other, older strikers, it was very different. They had families to feed and houses to pay for. We got strike pay at the end of the first week but none after that. It was to be a long, hard war of attrition.

Ma and Da were very supportive. I had a small amount of money saved, although not very much. I was 'a spendthrift', in Ma's words, and spent most of my wages on books and optics. However, ours was a trade union house and Da told me that I didn't need to worry about money, and that if I ran out of money, he would help out. In fact, from then on, he gave me a small amount of money each week 'to keep me going'. Paul also gave me a few quid during those first few weeks.

Being on strike gave me the feeling of being on an unexpected holiday. I would cycle out to Bull Island or Dún Laoghaire and spend all day discovering new species for me, taking notes and sketching the birds I was seeing. I was beginning to get to grips with new families of birds and learn the harder-to-identify groups like winter waders. On the last day of February, I even managed to see my first Red-necked Grebe. I had heard that one had been seen at Dún Laoghaire and I went out to see it for myself. I found it swimming inside the harbour of the West Pier. It was a superb bird. I realise now that this was the first time I ever went twitching. For the uninitiated, twitching is the act of going to a location with the specific aim of seeing a rare bird that has been found by someone else, so you can add that species to your list of bird species seen. Anyone who goes twitching is called a 'twitcher'. So I twitched that Red-necked Grebe.

March arrived and the strike continued. Each Monday morning I would meet my friends outside the GPO. We'd catch up as we did our picket duty and then we were finished for the week.

One morning in the second week of March, Paul arrived home early from work. I was not long out of bed and was surprised to see him. He informed me that the POWU had placed a picket on his depot and he was one of only a handful of men who did not pass the picket. This was a big worry for him, as his trade union would not support him. He could be in big trouble. He even risked losing his job.

'I don't care,' he stated. 'I will never pass a picket, especially when my own brother is part of that strike.'

I admired his courage and his principles.

When Da came home that night, we all sat down and discussed the situation. Now both Paul and I were not earning money. Da was very proud of Paul.

'It's one thing to be out on strike,' he said, 'but it's another thing altogether to be brave enough not to pass a picket.'

We had a united house and it was a wonderful feeling to be 'brothers in arms'. Ma and Da supported both of us as best they could, but we all had no idea that this would be a strike of endurance. In fact, it seemed that all of Ireland was in a state of revolution in March 1979. A big tax protest was organised and I marched along with about 100,000 other people through Dublin.

I was walking behind the POWU banner, and thousands of people cheered us strikers as we marched by. It was a powerful feeling to be part of such a protest. A few weeks later, I attended a POWU rally in the National Stadium on the South Circular Road. We then marched into the city centre and staged a sit-down protest in the middle of O'Connell Street. I was dragged off the street by the scruff of the neck by a big guard who gave me a few kicks in the back for my troubles. I carried my bruises with pride on the picket line the following Monday morning.

As March passed and April came, I was still out on strike and Paul was still not passing the picket. It was a tough time. Da was great in helping us out, but money was tight. During the second week of April, Paul's car broke down. He had a Hillman Hunter and it was his pride and joy. We, as a family, had never owned a car and Paul was the first in the family to have one. However, now, without enough money to get it fixed, it just sat parked outside the house. It soon became a source of great attention for one of the local characters, known in our house as 'Nedser'. He was a man who liked a pint and didn't miss an opportunity to make a few quid to keep him in drinking money.

One evening in late April, Paul looked out the front window to see Nedser and two other men standing around and looking at the car. As he watched, he saw one man kick the tyres and examine the bodywork very closely. They then all leaned on the car and seemed to be deep in conversation. Intrigued by their behaviour, Paul ventured out to see what was happening.

When he asked if he could help, one of the men said, 'Fuck off and keep your nose out of our business. We're buying this car off this gentleman.'

It was unbelievable. Nedser was selling Paul's car to these two men. They were surprised (to say the least) when Paul explained that the car was in fact his and it was not for sale. Such unwanted attention prompted Paul to sell the car shortly afterwards. He didn't have the funds to fix it, tax it or run it. The strike was taking its toll.

For me, while money was tight, time was bountiful. Besides my Monday morning picket duty, each and every day was spent out birdwatching. By now, I had learned a lot from being out and about and finding new birds all the time. I even learned that we went 'birding' and not birdwatching, and that I had 'bins' and not binoculars. I was beginning to speak and think like a birder. Spring came with Wheatears and Swallows, Whimbrels and Willow Warblers.

Spring that year was a time of great discovery and learning. At the IWC monthly meetings in Carroll's Theatre in Grand Canal Street, I met some of the greats of the Irish birding world. I met Redmond Wheeler, a grandfather of Irish ornithology. He was a real gentleman and was only too happy to share his knowledge with whippersnappers like myself. I would sit and listen to Colin Moore (C. C. Moore), one of the greats in Irish bird identification, as he spoke with the more senior birders about some complex bird identification feature he had noticed. It was a steep learning curve, but my young mind was soaking it all up. On the outings, I was with more experienced birders who shared their sightings with me. Equally, I was beginning to share my sightings with them. I also came into contact with other new, up-and-coming birders such as Brian Haslam, who I met when birding in Wicklow.

The last days of April seemed to drift unnoticed into the sunnier days of May and still the strike continued. I saw my first Roseate and Little Terns, Whitethroats and Spotted Flycatchers. I walked the coastal wetlands, saw breeding seabirds on the cliffs of Wicklow Head and hiked in the mountains to the sound of calling Cuckoos. As May slipped into June, it seemed there was a renewed effort to end the strike before the Government went on summer holidays. Negotiations were long and protracted. I watched the news bulletins with interest and dread. I really did not want this to end. I was enjoying my birding time too much. I vowed that one day I would bird for a living. For the moment, I just wanted to savour every single day in the field.

All good things do come to an end and I remember the moment very well. It was on RTÉ's *Six O'Clock News*. The newsreader,

Maurice O'Doherty, announced that the strike was over. The unions had secured most of their demands on pay and working conditions. It had been a long but worthwhile strike. We were to report back to work within days. I was devastated. I had hoped that negotiations would fail and that the strike would go on into the summer.

With a deep feeling of impending depression, I headed off with Don Conroy for my last day of 'strike birding'. It was a bright and sunny day. We headed into the Wicklow hills in search of Peregrine Falcons. I had found a pair a few weeks previously and it was a species Don wanted to see. The birds performed superbly well, with a juvenile bird even sitting out in the open on the ledge of a steep cliff. As the day wore on, we found ourselves at the waterfall at Powerscourt, and it was here we struck gold. I was walking in a semi-wooded area when I caught a glimpse of a bird that looked like a Wheatear. I raised my bins and took in the beauty of a male Redstart. The orange-red underparts, black throat and face, white forehead and slate-grey back were a shock to my system. I did not expect to see a Redstart, which is a very rare breeding species in Ireland. The bird took off, revealing that orange-red flash on the sides of the tail. It perched in a tree close to me, flicked its tail and called. I backed off. It flew again into a copse of trees only to reappear moments later. I called Don.

'I've just found a male Redstart!' I whispered to him.

We sat still and were rewarded with fantastic views of this beauty of a bird. Then, from the same copse of trees, a duller, browner bird flew into view and perched on a nearby rock. She flicked her orange-red tail. A female Redstart! We watched as the pair flew back and forth from the rock to the trees. It was obvious that they were nesting in the area – one of the very few nesting pairs in Ireland. We were delighted. I took notes and we both did field sketches. It was a great find.

When I got home that evening, I phoned Killian Mullarney to tell him that I had found a pair of Redstarts in Wicklow. He assumed it was a well-known pair that were present in a particular area of Wicklow and was more than surprised to realise that I had discovered a new, previously unknown breeding pair. That evening, birders were informed that Eric Dempsey had found a new pair of Redstarts.

It really wasn't such a major find. In fact, it was a very minor find, but for me in 1979 it felt like the moment when I first announced my presence to the birding family.

I went back to work the following day. It felt like I was going back to school after a very long summer holiday. I heard the following week that others had gone to see our Redstarts and that the pair had fledged two chicks from the nest.

The first letter I received when postal services resumed in late June was a Valentine's card that had been trapped in the backlog of post since February. I didn't know who had sent it to me. In truth, I didn't care. I was already in love … with birding.

Chapter 11

MAKING THE GRADE

During the summer of 1979 I spent most weekends out birding. On 13 July, two days before my eighteenth birthday, I saw my first Barn Owl with Don. It was at Donadea Castle in Co. Kildare. It was a sight I had imagined since childhood and there it was, a ghostly shape soaring down from the darkness of the castle. It was the highlight of the summer and was one of the best early birthday presents I have ever received.

Being back at work after the strike took some adjustment. Once I turned eighteen, I should have been assigned to the position of Postman, but somehow my personal file went missing from the Registry Office, where all such files were kept. I am not admitting to anything, but it was well known that if a JP wasn't keen on going to Sheriff Street then their appointment could be delayed if their file went missing. Let's just say that some of the clerical staff I knew could not find my file when they went to look for it. So, I would have to remain on as a JP until the file could be located. This suited me perfectly as I did not want to be consigned to 5.30 a.m. starts, weekend shifts or night work, especially with the autumn approaching.

As the first waders began to return, I felt a twinge of excitement at the thought of the autumn ahead. For any birder, the autumn is *the* time to see new and exciting birds. In autumn, birds are on the move. From late August to the end of October is the peak autumn migration period. Most of them are young birds, migrating for the first time. Individual birds get lost, go in the wrong direction or get blown off course in storms. This is when rare vagrants from Siberia touch down in Ireland. It is when North American vagrants can be spotted on our shores. This was going to be my first 'real' autumn as a birder, an intense and exciting period in my young birding life. I was primed.

So, August saw me cycling out to Bull Island most evenings after work and spending hours looking through the flocks of waders as they fed on the mudflats. 'Through' is perhaps an unusual word to use, but that's exactly what you do. You look *through* the flocks of common birds in the hope of finding something rare or unusual with them. It's like looking through a crowd of people to pick out someone you know.

More importantly, I was now on the periphery of 'the grapevine'. This was as important as having a pair of bins or a scope. The grapevine worked on a simple basis. If you found a rare bird, you would phone several birders who were close to you. They in turn would phone others who would, in turn, phone another bunch of birders and so on. So, being part of the grapevine meant that your telephone number was with other birders and, should any bird of note be found, you would get a phone call to alert you. The inner sanctum of the grapevine was still a long way off, but for now news of sightings reached me within twenty-four hours.

I was birding with a gang of other guys around my own age and getting to grips with autumn-plumaged Dunlin, Curlew Sandpipers, Little Stints and Green Sandpipers. I saw Ruffs and Reeves for the first time and Brian Haslam showed me my first Black Tern at Swords on 25 August. I also went seawatching for the first time. This was a new and wonderful experience. I saw rafts of Manx Shearwaters and my first Great and Arctic Skuas off Wicklow Head.

It was the grapevine that presented me with my first Yankee wader (to be more correct, they are North American waders, since many breed in northern Canada). Kieran Grace ('Gracer') was a young birder who lived in Clontarf and it was he who broke the news to me that a Wilson's Phalarope was 'showing well' at the Swords Estuary. I learned that the words 'showing well' (as in the bird was giving great views) were enough to fill me with excitement and adrenalin for many years to come. I rushed out to Swords that evening where most of the top twitchers had gathered. This beauty of a bird with its long, delicate needle-like bill, long mustard-yellow legs and pale grey upperparts was the embodiment of what a rare bird should look like.

Among the birders gathered there watching the 'Wilson's Phal' were Jim Fitzharris ('Fitzer') and Jim Dowdall. They were a few years older than me and were associates of the Killian Mullarney ('Muller') group. Among us slightly younger birders, these guys were legendary. They were sharp birders and, rumour had it, they did not suffer fools gladly. If I am honest, they were quite intimidating for many of us. The two Jims had made their name by finding Ireland's first Rock Thrush, a very rare European species, at Clogher Head, Co. Louth, in May 1974. Finding a 'first Irish' was enough to catapult these two young birders to the fore of Irish birding. Since then, they had proved themselves worthy of such acclaim, as they were (and still are) top birders. I listened in on the banter and the general bird identification discussion among these lads. I was still on the edge of this group and felt in my heart that I would need to prove myself to be accepted by them.

For my birthday that summer, Ma and Da had paid for a subscription to *British Birds*. This was a monthly journal packed with ground-breaking identification papers and write-ups of various findings of rare birds. It also had a 'Recent Reports' section where I could read about the rich variety of rare birds that had been seen in Britain and Ireland. It was enough to whet my appetite for all sorts of possible vagrants that I might see in the coming autumn. I waited anxiously each month for my copy of *BB* to arrive and would sit for hours reading the minutiae on the coverts of Semipalmated Sandpipers or the identification of Western Palearctic gulls. It was the bible of birding.

The editor of *BB* was J. T. R. Sharrock. He was one of the founders of Cape Clear Bird Observatory on Cape Clear Island off the coast of Co. Cork. The island's position in the Atlantic Ocean off the south-west coast, and its proximity to the Fastnet Rock lighthouse, means it is a magnet for migrant birds in spring and autumn (many birds migrate at night so are drawn to islands and headlands with lighthouses). So, in autumn 1959, Sharrock and others visited Cape Clear and began to document what they saw. They found incredible birds, and from that year on the island became the autumn Mecca for Irish and British birders. Alas, the famous Cape Clear Bird Observatory is now closed.

I had picked up a copy of a book called *The Natural History of Cape Clear Island*. It was written by Sharrock and illustrated by the great bird artist Robert Gilmore. I read it from cover to cover many times. It documented many great birds – birds I hadn't known even reached Ireland. It told tales of days when American and Asian warblers were seen side by side. My imagination ran riot.

Locations on Cape Clear have legendary status in the lore of Irish birding. There is Cotters Garden, the Post Office Garden, the Waist, the Escallonia Garden, the Youth Hostel Garden, West Bog, East Bog, Central Bog, the Wheatear Field, Lough Errul, the Secret Valley and the High Road, to name but a few. Each location has hosted many rare and wonderful vagrants. Then, of course, there is Blananarragaun.

'Blanan' is a long, narrow, treacherous headland that sticks out like a stony finger into the Atlantic Ocean. It is the most southern point of Ireland. It was here that, when strong autumn storms struck Ireland, hardened birders risked their lives climbing over the notorious blowhole before clambering across the sharp, slippery rocks and out to the tip, ignoring the sheer drop to the right. Being perched on the most southern tip of Ireland in dreadful weather conditions, seabirds passing within feet of you, was worth every risk to these hard-core birders.

There was no doubt that Cape Clear was the place to be in autumn, and in September 1979 that was exactly where I chose to go. I went with my cousin Fergus, who was an early birding companion. I had booked us into the Bird Observatory. He was staying for just a few days; I was booked in for two full weeks.

Getting from Dublin to Cape Clear Island in 1979 required planning of epic proportions. We travelled down to Cork on the train. We spent the first night in Cork, eating a healthy dinner of fish and chips, having a pint or two, and pondering on the days ahead. The following morning, we caught the first bus to Skibbereen. It was vital to get that first bus because that gave us enough time to catch the Skibbereen to Baltimore bus. If you missed that bus, then you'd miss the 2.15 p.m. sailing to Cape Clear.

And so it was that, on 19 September 1979, I first stepped on board the ferry the *Naomh Ciarán* and set sail out into Roaringwater Bay. It was a choppy journey. As we rounded the small islands close to the shore and set out to sea, the magical misty outline of Cape Clear gradually became clearer and soon we were alongside the rocky shoreline, turning in to the picturesque North Harbour. From the boat I could see the hill that runs up from North Harbour and there, on the hill, was the famous Cotters Garden. The Bird Obs was also right on the harbour.

The warden of the Bird Obs was an elderly man called Colin Rhind. He was a true gentleman and spoke with a very rich, well-pronounced posh English accent. He welcomed us to the island and brought us to the Obs. I walked into the small kitchen where there were one or two other young birders. One, Michael O'Donnell, was a Dublin birder, and over the course of the following day or two he showed us the main gardens and areas to watch for migrants. Most importantly, he also brought us out to Blanan and taught us the best and safest route to take.

From the small kitchen, a door led into the main room, which contained books and the important logbook of sightings. I took in some of the pictures on the wall. There was an original pen-and-ink drawing of a Rose-breasted Grosbeak by Robert Gilmore from J. T. R. Sharrock's book on Cape Clear. The sense of birding history oozed from the walls. How many great birders had sat around this table and recounted tales of rare birds found over the years?

Off the entrance hall was another room, where boots, jackets and bins were kept. It smelt of dirty, wet, greasy Barbour jackets. This odour still brings me back to that room. Up the narrow stairs were the bedrooms. There was a room at either end of the narrow landing that housed four people in each. The bunk beds were old and the mattresses were thin and lumpy, to put it mildly. One other room had bunks where two could sleep. Then there was the wardens' room. Around the left corner of the house was a small building with a large chemical toilet. Around the right corner was the washroom with a sink and a makeshift shower. It was primitive, but this would be home for the next two weeks.

The Bird Obs was run like a hostel and we all had duties to carry out during our stay. One of my first duties was to empty the contents of the chemical toilet down a drain near the harbour. It was a disgusting job, but it seemed it was a rite of passage for all newcomers and I accepted it with grace. We paid a small nightly fee to stay in the Obs and everyone cooked for himself or herself. I came armed with tins of beans and sardines as well as a fortnight's supply of Smash, the powdered mash potato mix. I was there for birding, not for gourmet cooking. I had a big box of cornflakes for breakfast and bought a big carton of long-life milk that tasted horrible (fresh milk was hard to get on the island). There were few comforts to be had.

I set off to stand in the famous Cotters Garden, watching my first Choughs soar across the harbour and over the hill. Cotters Hill was steep but was nothing compared to the 'A1', which ran at a 45-degree angle from the harbour to the north end of the island. The island was full of bangers of cars and it was incredible to watch old Beetles struggling up this sheer cliff of a road. I sat in Cotters Garden and watched for migrants, but all I saw were Robins.

That night, I had my first experience of the evening rituals of life in the Obs. After we finished cooking and eating, the radio was switched on. It was time for the shipping forecast on BBC radio. Everyone sat listening intently to the meteorological situation over the British Isles, with fast-moving depressions south-west of Iceland or large anti-cyclonic pressure systems off to the east. Then coastal reports from all the lighthouses were read and we waited for the report from Fastnet Rock lighthouse, which was perched precariously on the bleak rock to the south of Cape Clear. These reports gave us all the information we needed to judge what the weather would be like and, most importantly, where the winds were coming from. Southerly winds might just pick up southern migrants. Strong westerly winds might blow in an American warbler. Listening to the shipping forecast allowed us to form a mental picture of the weather systems and, in turn, to try and predict what birds might be in Cotters the following morning.

It took a while for me to understand the complexities of the weather systems described in these forecasts, but gradually I grasped the basic concepts in terms of migrating birds. I was soon joining in the disappointed sighs when north-westerly winds were forecast and smiling with anticipation when southerly or easterly winds were predicted. It was a superb education to get from the more seasoned birders.

After the shipping forecast came the daily log. A list of species was read out and we each contributed our list of sightings and counts of the different bird species seen so that a full daily picture of the birds on the island could be documented. These reports were meticulously scribed by the warden into the logbook. It was only then that everyone could adjourn to Club Chléire. The Club was a barn-like pub in North Harbour and the only pub open at night. There we would sit and discuss the day and what tomorrow might bring. There were never any long nights in the Club. Each morning, we would rise before dawn and be out birding at first light.

Over the next few days I walked the length and breadth of the island. I met many of the islanders, including Dinny Burke who ran the small shop at the top of Cotters Hill, as well as his cousin Paddy Burke who had a very small pub/shop also near the top of Cotters Hill. Paddy was a tall, round, happy man, with red cheeks and a big smile. He liked nothing more than regaling us with stories of when Tim (Sharrock) first arrived on the island in 1959 and how he went birdwatching with him. He particularly enjoyed telling us all of the famous Bee-eater (said with a wonderful Cape Clear accent) he had seen that autumn.

I also met Mary Mac who lived near Lough Errul. I had heard that she baked bread and so I knocked on her door to see if I could order a loaf. We chatted briefly and I ordered scones and soda bread. She told me I was more than welcome to cross any part of her land. Her bread and scones kept me going for days. From then on, she always greeted me with the words 'Hello, Eric Boy, how are you? Any good birds around?'

Her land led down to the West Bog from where the stark outline of the Fastnet Rock lighthouse could be seen out to sea. It was this light that attracted all those night-flying migrants and I felt that the West Bog had to be one of the first places they would land. Despite my best efforts, however, there was hardly a migrant to be seen. I was learning the hard way that all of those reports of rarities were just that: they told of rarities among the millions of Robins, Dunnocks and Stonechats that seemed to be in every square metre of the island. The best I saw over the first five days was a few Spotted Flycatchers, Wheatears, Whitethroats and a Garden Warbler or two. I had also been out on the tip of Blanan and seen my first Sooty Shearwaters, birds of the southern oceans. However, after five days, I was beginning to feel that Cape Clear Island was not living up to its reputation.

On 24 September, Fergus left the island. I waved him off on the morning boat and walked up to Cotters. I instantly saw a couple of Willow Warblers and a Chiffchaff. It is always a warm feeling to know that migrants had arrived. I checked the Waist and saw a

Garden Warbler. Near the Youth Hostel I found a female Redstart. I walked the entire north end without seeing anything else of note. It seemed that Cape Clear was toying with me, giving me a taste of what could be. I returned to the Obs and met two new birders who had arrived on the afternoon boat. They were Tony Lancaster and his partner, Pat Hamilton. Tony was a Cape Clear legend in his own right, having been to Cape Clear with the likes of Sharrock in the early 1960s. He was part of the group that had found and identified Ireland's first Bonelli's Warbler in 1961, the year that I was born. Tony and Pat were very pleasant company and were happy to share their knowledge of both the island and the birds.

Later that afternoon, whilst walking near East Bog, I met Pat. She had just found a Pied Flycatcher. It was a new species for me. I enjoyed watching this beautiful bird flying up from the ferns to catch midges. It felt as if Cape Clear was beginning to hot up.

The following day, 25 September, dawned wet and wild. Strong southerly winds, gusting to storm force, along with prolonged periods of heavy rain, battered the coast. It was an ideal morning to risk our lives and head out to Blanan. In near darkness, we made our way out to the tip. The seabird passage was superb, with lots of Sooty Shearwaters and Great Skuas. The best bird for me, my first Pomarine Skua, passed so close to where we were sitting I could almost reach out and touch it. Its deep chest and powerful, broad wings made it a totally different beast from Arctic Skua. I was well pleased with my morning's work when I gingerly retraced my steps off Blanan and returned to the Obs for a much-needed bite of food.

I remember the moment when I first saw the figure running down Cotters Hill as if it were yesterday. I was in the kitchen and had beans cooking in a pot and bread toasting under the grill (the delights of beans on toast after a seawatch can't be described). The figure looked like Tony Lancaster. There is only one reason why a birder runs on Cape Clear and that is when he or she has found a major rare bird. Tony was galloping down the hill towards the Obs. I turned off the gas cooker, shouted to Michael who was upstairs, grabbed my bins and ran. We met Tony at the harbour. He was panting.

'Subalpine in West Bog!' he said through deep gasps. 'It's in the gully at the top end.'

I had to think long and hard for a moment. What was a Subalpine? I found myself mentally flicking through my trusty *Peterson Field Guide* … Subalpine? Subalpine?

'Subalpine Warbler?' I blurted out.

'Yes … Subalpine Warbler … West Bog!'

Before he had finished the sentence, we were already running. I had remembered seeing the words 'Subalpine Warbler' in the field guide but I hadn't a clue what one looked like. In birding parlance, really rare birds are known as 'megas'. This was a mega and I needed to see it. We ran up Cotters Hill in seconds, breaking Olympic sprint records. We continued on past Paddy Burke's, up the deceptively steep High Road, around the edge of Lough Errul and past Mary Mac's.

'Hello, Eric Boy. Any good birds around?' I heard a voice ask.

'Yes, a Subalpine Warbler!' I answered as I ran past her house down towards West Bog. I said it as if she knew what a Subalpine Warbler might be.

'I hope ye see it,' I heard her call after us as we disappeared down the lane.

So do I, I thought to myself as I sloshed through pools of water.

Pat was at the far end of West Bog. She was standing overlooking the small gully where a stream emptied into the cove below. We ran up to her. The bird was skulking in the vegetation and bushes around the gully but had not been seen for over twenty minutes. My heart sank. Was it gone? Where might it be now? Then I saw a movement at the edge of the ferns and caught a glimpse of a bird. Its upperparts looked to be a bluish-grey. It went back into cover. The rustle of vegetation marked its progress through the undergrowth. It appeared again and sat perched out in the open. It was a thing of beauty: a bluish-grey crown, nape and mantle, a red eye-ring, and pinkish-orange underparts with striking white moustachial stripes running from the bill down below the cheeks on either side. It perched upright and flicked its white-sided dark tail before going back into cover. It was like a small, colourful, exotic Whitethroat. We stood in silent reverence for one of the best birds we had ever seen in our lives. It 'screamed' rarity. I took in my surroundings. The vista of the Fastnet Rock lighthouse was ahead of me out to sea, the West Bog behind me and the Wheatear Field running up the hill to my left. I was on Cape Clear and had just seen a Subalpine Warbler. From that moment on, West Bog became my favourite place on the island.

The warbler showed well on and off around the gully for the afternoon. We left it skulking in the thick vegetation and returned to the Obs for our much-needed food and to write our detailed notes of our observations into the Bird Obs's rarity-description log. The bird was not seen the following day, but a concerted effort by a fresh arrival of new birders on 27 September was successful in locating the bird in the same general area. Later that day we heard that an immature Red-backed Shrike was seen in the Secret Valley, near the church at the north end of the island. Several of us ran up the A1 in fading light, but despite our best efforts the bird could not be found. I have to admit that this was probably the first and last time I ever ran up the A1.

The following morning had me up in the Secret Valley at dawn, watching a fantastic Red-backed Shrike devouring a large beetle. It was a young bird with crescent-shaped barring on the underparts and across the rusty red upperparts. Cape Clear wasn't just living up to its reputation but was surpassing it beyond all my wildest dreams.

I took the morning off the island and went birding with Tony and Pat on the mainland. We headed to Lissagriffin, near Mizen Head, where we found my second Yankee wader, a Pectoral Sandpiper. It was so tame that it walked within a metre of us. It was a wonderful morning and we returned to Baltimore in time to catch the 2.15 p.m. sailing back to Cape Clear. Other birders were arriving, including the well-known British birder Tony Marr. He was very tall and lean and had huge bins and a Barbour jacket that was several sizes too small for him. He was a real gentleman.

The following day saw the arrival of more migrants, including my first Yellow Wagtails and Firecrests, as well as two Redstarts and a Pied Fly (by now I was speaking 'birder' fluently and was calling this species Pied Fly as opposed to Pied Flycatcher). Everywhere you went there was a feeling that something 'big' might just reveal itself. It was this feeling of the possibility of finding a mega bird that kept us all going for hours each day.

However, of all the days of that trip to Cape Clear, 30 September stood out as a very special day for me. It was a dreadful day with prolonged heavy rain for the whole morning and into the afternoon. There was a strong south-easterly wind. That afternoon, a whole bunch of birders were due to arrive to stay at the Obs. These included Gracer and Jim Dowdall. The Obs would be full. I headed out early. There seemed to be lots of common migrants on the island. Returning back to the Obs, I met Gracer and others walking up Cotters Hill. Gracer stood and chatted whilst the others nodded in my direction and continued on. He congratulated me on having had such a great week, especially on seeing the Subalpine Warbler, which was only the seventh record of the species in Ireland.

I went back down to the Obs and had some lunch. It was still lashing rain and I had my jacket hanging to dry. I was enjoying my beans on toast when suddenly in burst a very sodden Dublin birder by the name of Dermot Hughes. His glasses were steamed up and he was soaked.

'Short-toed Lark at the lake!' he shouted.

Plates, cups and books went flying as people grabbed jackets, bins and scopes. We ran up to Lough Errul where this dumpy little lark was feeding alongside a Yellow Wagtail on the short grass on the northern shore. Like us, it was soaking wet, but as the rain eased off, the sandy colour of its plumage revealed itself. It was yet another mega bird for me (and everyone else). I stayed watching the bird while some of the others headed off to West Bog. With so many migrants on the island and the rain now stopping, there had to be other great birds to find.

I followed on to West Bog about an hour later and was met by Gracer, Jim Dowdall and some others. They had just found a Lesser Whitethroat in deep cover. This was yet another new bird for me. They showed me where it was last seen but warned that it was skulking and that they had only seen it briefly on a couple of occasions. I headed down to the large area of bushes and ferns where they'd seen the bird. I stood still and waited. I saw a movement. The sun came out briefly. The bird came out briefly. I raised my bins. The bird was facing me. I saw a red eye-ring, white moustachial stripes and a pinkish-orange breast. This was no Lesser Whitethroat – this was the Subalpine Warbler that had not been seen for days.

The bird dropped back into cover. I panicked. Was I sure that I had seen what I thought I'd seen? What if I called all of those fellas back and it turned out to be a Lesser Whitethroat? I took a deep breath and reassured myself that it had to be the Subalpine. I glanced over beyond West Bog. The others were not far away. I hesitated again. What if I was wrong? My reputation would be in shreds if I were mistaken. I thought for a few seconds more, took another deep breath and then ran after them, catching their attention with a sharp whistle. They came towards me.

'It's not a Lesser Whitethroat: it's the Subalpine!' I announced.

Jim Dowdall looked at me. 'Are you sure?'

'Yes, 100 per cent,' I answered with a confidence that belied the fact that inside I was now only 10 per cent sure and beginning to regret ever running after them.

We all returned to where my scope stood guard. We waited. Nothing moved. It seemed like an hour before a movement was picked up low in the bush. The bird then emerged – it was the Subalpine. It was a new bird (a 'tick', named after the old habit of ticking off the birds you saw in the back of your field guide) for everyone. I felt a surge of relief through my entire body and, weakened by the experience, left them to enjoy this wonderful bird. Out of view, I sat down on a rock and let out a long deep sigh.

'Thanks be to Jaysus it was the Subalpine,' I found myself saying to a farm dog that came up for a pet.

That night I joined the gang of birders in the Club for a drink. I went to the bar and ordered a pint. Everyone got at least one tick that day. For some of the new arrivals, the Subalpine meant they had seen two new species. I sat down and supped my pint, enjoying the chat and banter. Jim Dowdall appeared in front of me. He placed another pint on the table before me.

'Great bird, that Subalpine,' was all he said.

He didn't need to say any more. For me, that day marked a coming of age, my rite of passage. I felt accepted into the tribe of birders. It was one of the best pints I had ever tasted in my young birding life. It is odd to write the words 'birding life' as opposed to simply 'my life'. The truth is, from that day on, my birding life and my life became intertwined and will remain so for as long as I live.

My last day on Cape Clear was another memorable day, with a Wryneck showing well on a wall in West Bog. This cryptically plumaged member of the woodpecker family crawled along the wall, feeding on insects with its long anteater-like tongue. It was a fitting end to an incredible second week. I had seen a Subalpine Warbler, Firecrest, Red-backed Shrike, Short-toed Lark and Wryneck, to name the rarest of the birds.

On 2 October, I left the island on the morning sailing. Several birders waved me off. Colin Rhind waved me off. As I rounded North Harbour and sailed out into Roaringwater Bay, I watched Cotters Garden slipping out of view. I vowed I would return. I had Cape Clear Island in my birding veins. I knew that I was now a birder for life.

Chapter 12

BECOMING A RUNNER

The autumn of 1979 passed with few other birds of note. I was out each and every weekend birding at the main sites of Cos. Dublin and Wicklow with a new determination and drive. Of course, having enjoyed the highs of Cape Clear, I once again resigned myself to hours of scouring migrantless headlands in the hope of finding the big one. However, I was not complaining. The autumn of 1979 marked a breakthrough for me not only as a birder but also for my 'Irish list'. The Irish list is the key list and is a tally of all the species you have seen in Ireland.

On Cape Clear I learned that you were not only judged on your ability to find birds but also on what your Irish list was. This was a much-asked question and was proof of your vintage and experience. I had heard that the Cork birder Ken Preston had a list of around 264 species. As a twitcher, your aim is to see as many birds as possible so that your list is as high as possible.

In 1979, I thought it incredible that anyone could have seen as many as 264 species of bird in Ireland in a lifetime. Today, that seems like a lowly list for a top birder. In fact, in recent times, some birders have seen almost that many species in Ireland in a single year. In 2003, I saw 230 species in Ireland and that was just from being out birding. I wasn't even trying to get a big 'year list' (the list of bird species you see in a year). With mobile phones, texts, the internet and so on, news travels so much faster. But back then, news from Cape Clear was often released to Dublin by letter. Now, when a rare bird is found, people hear the news within minutes and can look at their phones to see digital images immediately. I sound like a real old codger writing this, but because of modern technology, better cars and better roads, twitching is so much easier these days. Just think how much faster it is to get from Dublin to Cork now, in comparison to 1979. I have often been from Dublin to Cape Clear and back in a single day, whereas back then it took me two days to get there. So, twitchers get news faster and can travel faster than ever before. As a result, several birders now have Irish lists of well over 400 species.

It is important to say two things about twitching in Ireland. First, we were, and still are, birders first and twitchers second. Days are spent in the field birding favourite habitats in the hope of finding that rare bird that others will twitch. Many of the top Irish twitchers might only travel to see new birds four or five times each year, but that doesn't stop them from being out every weekend looking at birds. Many also serve their time with conservation bodies, making enormous contributions to the protection of Ireland's birdlife and habitats. In Britain, there are tens of thousands of twitchers who only don their bins when there is a new rare bird to see. That has never been the case in Ireland.

Second, there is a danger, when speaking of lists, that twitching can be reduced to simply collecting ticks in the same manner as trainspotters might collect train numbers. However, twitching and getting a tick is so much more than just collecting numbers. In late 1979, when I jumped headfirst into my twitching life, there was a great feeling of being on the cutting edge of birding. Twitching required travelling long distances with some of the top birders in the country. There was a sense of adventure and discovery. In truth, there was also a slight air of elitism. Only those within the inner sanctum of the grapevine were on the road. During the hours of driving, complex identification features and concepts were discussed at length. For a young birder, it was an education of a lifetime. I learned more from other birders during twitches than I could have learned in a thousand books. I also learned from seeing and experiencing birds first hand. These discussions were also important for keeping the driver awake as we raced through the night to be at some remote headland in Co. Cork or Co. Clare for dawn. There was always the inevitable talk of the weather. Cloudy, foggy or wet nights always bode well for finding the bird, as most night migrants are reluctant to fly in such conditions. Clear, starry nights fill birders with dread: starry nights allow birds to navigate well and often sound the death knell of our dreams.

For a young birder like me, there was also the sense of great fun and a feeling of camaraderie from just being out with birding characters like Boss Cummins, the Commander, Big Willy, Chick, The Bag, J. C., the Kid, Bomber, Ludigi and Dipper, as well as Dowdall, Fitzer, Gracer, Muller and McGeehan (to name but a few).

Then, once we got to the location where the bird had been seen, there was the skill required to actually get out and find it. It might have been a warbler last seen in a hedge in the middle of a large headland. There could have been miles of hedgerows. We'd all spread out across the headland in search of our bird. Only good birders can locate such birds, and, to be part of the birding tribe back then, you had to prove yourself worthy of being on the twitch. If someone found the bird, then people would come running across fields, clearing open ditches and hedges to get to where it was. The relief of seeing that bird was palpable. You got your tick. You got your bird. Hours would then be spent watching it, learning the key features, taking in the way it moved and behaved, photographing it or sketching it, and taking notes. The whole experience was intense.

The high of successfully 'connecting' with a rare bird is incredibly addictive. It could keep you going for weeks. If it is a mega bird, it might keep you going for months. The species is added to your Irish list. If it is a bird you have not seen anywhere before, then it is also a 'lifer', a species that is added to your life list (for those who go birding abroad). For birders who are keeping a year list, that bird also becomes a superb 'year tick'.

On the reverse side, the low of missing the bird (or 'dipping', as it is known) is equally intense. Dipping is such a low that it is hard to take at times. Dipping on a mega bird can be agonisingly hard to take. I have known birders who plunged into the depths of depression when they dipped on a mega bird. Regardless of whether you connect or dip, the twitch and the bird become another part of Irish birding lore. By 1979, I was now a part of this lore.

Twitching was the adventurous and exciting side to birding, and this was the magical and mad world in which I, as a young eighteen-year-old birder, was about to immerse myself. Leaving Cape Clear in 1979, I was a long way from that level of birding intensity. I did have an understanding of what twitching was all about: dashing from the Bird Obs to see the Subalpine Warbler or Short-toed Lark was a form of twitching. The high of seeing those birds was infectious. I listened to the other lads speak of famous twitches and realised that the real difference was that on Cape Clear Island I might dash two miles to see a bird, while they often dashed 200 miles to see one. It was all a matter of distance and the ability to get there quickly. I returned home that October and immediately set about counting up my Irish list. I had no idea how many species I had seen. I read through my notebooks and did my calculations. As of 3 October 1979, my Irish list stood at 161, almost 100 fewer than Ken Preston's. So, I set myself a task. I would strive to reach 200 by the end of 1980. Back then, an Irish list of 200 species was seriously respectable.

There was one major problem, of course: I did not have a car. So, I was restricted to seeing birds that were within a bus journey from Dublin. IWC outings were the exception, since everyone carpooled and you got a lift. On one such trip in early November, I saw my first Velvet Scoters off Gormanstown, Co. Meath. Muller was leading the trip and picked up a small group of them flying within the Common Scoter flocks. I got onto them straight away, noting their startling white wing patches against their all-black body. I was one bird closer to my goal, but by now I realised that getting there was going to be painfully slow.

However, my life changed with a phone call from Gracer. He was heading off birding to Co. Wexford with Ronan Hurley and there was space for a 'runner' in his car. I didn't know Ronan well. He was an active birder and a frequent twitcher. He had also just been made chairman of the Dublin branch of IWC. His giving me the opportunity to be a runner was an answer to my problem.

A runner was a passenger in a car who paid a share of the petrol costs (known in birding as 'sting'). At the end of the day, the driver 'hit their runners for sting' (asked their passengers to contribute their share towards the petrol costs). In the early days of birding, few people had cars. Without a car you simply couldn't go birding to places like Wexford or Cork very easily. Without a car, you couldn't go twitching. However, if you got a chance to be a runner, it was the ideal solution. However, a runner was much more than just someone who helped with the cost of petrol. Runners had to be good birders too. A runner was expected to contribute to the birding and twitching by being able to find birds. A runner also had to be tolerable company. I saw runners being thrown out of a car in Cork and told to get the train home to Dublin, after having annoyed their driver. If you got that kind of reputation, then nobody would want you as their runner. However, if you got established as a runner in a car, then you were guaranteed first refusal on any twitch.

So, on Sunday, 11 November 1979, I entered the world of the runner. It was very early when Gracer and Ronan turned up at my house in Ronan's tiny red two-door Fiat 127. I sat in the back. The car was just about big enough to carry three people plus scopes, tripods, bags and boots. This was the first time I travelled in this famous Fiat 127, and over the following years it transported me to the four corners of Ireland and back again. I often wonder just how many miles I travelled in that car. I slept in it countless times. It was the cocoon for many wonderful debates on everything from wing coverts to the existence of God. This little car was such a part of my early birding that my mind is filled with too many memorable journeys to recount them all here.

The day in Co. Wexford was fantastic, despite the cloudy, wet weather. We drove around the Slobs (large expanses of reclaimed costal grasslands), looking at geese from the car. It was a new experience. On the North Slob, I saw my first wild Canada Goose among the thousands of Greenland White-fronted Geese. From here, we drove over to the South Slob, and, from the wall overlooking Wexford Harbour, we encountered six Black-necked Grebes and a Slavonian Grebe. It could not have got better than this. Without a car, this level of birding was impossible.

Ronan was a gentle, quiet-spoken and highly intelligent fella who held firm and strong views on many subjects. Like Gracer, he too was from Clontarf and was about five years older than myself. He had been birding for many years but was as good a botanist as he was a birder. Both he and Gracer were members of the Dublin Naturalist Field Club, one of the oldest of such clubs in Ireland. He was a senior civil servant at a young age. I was also officially a civil servant as I had recently passed the Clerical Assistant (CA) exam and interview. This was the lowest grade in the civil service ranks, but it was a big step up from Junior Postman. It also meant that I had avoided promotion to Postman and so never had to step foot in the dungeon of Sheriff Street, the main sorting office. Incidentally, once I had received notice that I had been successful in the CA exam, my file was mysteriously found by my clerical friends in the Registry Office in the GPO (how lucky was that?).

Ronan was also an established and talented photographer. He was, however, also known for being a dreadful driver. He had never sat his driving test. He secured his full driving licence as part of a famous amnesty handed out in the late 1970s to all those on their third provisional licence. One birder recently recounted that sitting in the car with Ronan was the only time that he became religious. He would sit in the back seat with a set of rosary beads, quietly reciting novenas until he was safely home. This is perhaps a case of classic birder wit and exaggeration, but let me put it this way: Ronan's driving always kept me alert.

During the course of our day in Wexford, Ronan and I got on famously, and by the time we got back into Dublin he had invited me to join the committee of the Dublin branch of IWC. At the AGM in December 1979, I was voted on. I was apparently the youngest serving committee member of the branch ever. It was the start of over thirty years of uninterrupted voluntary service with IWC/BWI branches.

Ronan became a close friend of mine and was always considered a very welcome part of my family. He would sit for hours and join in heated debates with Da about politics, religion and everything else in between. However, in late 1979, the important fact was that on the return leg of that first Wexford trip he invited me to become a permanent runner in his car. The world of birding and twitching had now opened up and that goal of 200 species began to seem realistic.

Chapter 13 ⤙

THE BIG 200

S o, I was on a mission to see 200 species in Ireland and 1980 was the year in which I planned to achieve it. The winter of 1979–80 was one of local birding. It seemed that the best birds were along the east coast. I was birding at each and every opportunity available to me.

It was a time of plenty for a young birder. In November, I saw my first Glaucous Gull in Skerries, north Co. Dublin, while out looking for a Mediterranean Gull, which had been seen there the previous day by Brian Haslam. Med Gull was a major rarity back in 1979. Nobody could foresee that this species would expand so far out of Europe that they would become quite a common winter visitor to Ireland.

The year 1980 began with a real flurry of activity when, on New Year's Day, I joined a big gang of birders out at Rogerstown, north Co. Dublin, in search of Ireland's fourth Black Duck ever. This rare North American duck had been found by another young up-and-coming birder called Hugh Gallagher. We didn't see the bird that day, but I enjoyed distant views of it when it was re-found on 5 January. A week later, I was watching my first Black-throated Diver at Dún Laoghaire. The year was off to a great start and my goal of seeing my 200th Irish bird by the end of the year was beginning to become a real possibility.

In late February, I found myself in esteemed company when I was invited to become a runner with none other than Dowdall and Fitzer on a twitch to see a possible White-billed Diver in Youghal, Co. Cork. It was almost a year to the day since I had gone on strike, and here I was – a runner with Fitzer and Dowdall. We gathered along the seafront with every other birder in Ireland and looked for that diver. In the end, it was decided that a Great Northern Diver, with a very pale bill, was most likely the bird that had been mistaken for its rarer, more northern cousin.

As winter drifted into spring, I went to Co. Donegal for the first time. It was the St Patrick's weekend. I was travelling with Ronan, Gracer and Micheál O'Brien. Micheál had won the Young Scientist of the Year Award in 1977 for his studies of Brent Geese in Dublin. It was my first long-haul trip in the red Fiat and, with four birders and our gear, it was a tight squeeze. We were birding the harbours and fishing ports in search of seaducks and gulls. I saw my first Eiders, Barnacle Geese and Iceland Gulls. I was clocking up the ticks.

Co. Wicklow served up new birds in spring and summer: Crossbills, Grey Partridges, Red Grouse, Wood Warbler and, in the valley of Glenmalure, my first Ring Ouzels. This was once a traditional place to find these elusive and shy Blackbird-like birds. Their plaintive songs seem to reflect the vastness and remoteness of the valleys they inhabit. Ring Ouzels are no longer present in Co. Wicklow and are now a very rare Irish breeding species.

By the time the summer ended and the autumn commenced for real, I was primed to make a full assault on my birding ambitions. I had booked a two-week holiday on Cape Clear Island and I was a runner with Ronan. This was to be *the* autumn when my ambition of that magical 200 would be fulfilled.

September 1980 could not have got off to a better start. On 1 September, I had just arrived home from work when there was a phone call. It was Gracer.

'Killdeer on the Bull!'

I grabbed my birding gear, asked Ma to put my dinner in the oven, got back on my bike and cycled as fast as I could to the North Bull Island. It's a good 15 km at least but I believe that I could possibly have won the time trial in the Tour de France that evening. Brian Haslam was on a roll. He had found this exotic North American 'double-ringed' plover on the Aldermarsh. It was a magnificent find. Gracer kindly gave me (and my bike) a very welcome lift home in his dad's car.

The following week, I was 'running' with Ronan down to Co. Wexford. On 7 September I was given an 'honours lesson' in wader identification at Tacumshin Lake by some of Ireland's elite birders as they talked me through the subtle differences between juvenile Little Stint and juvenile Semipalmated Sandpiper (Semi-p). Little Stints are common European passage migrants that occur each autumn in Ireland. Semi-ps are their rare North American cousins. Its name originates from the fact that it has partially webbed feet, 'palmation' being the correct term for webbing. This bird was among the stints and was the first Semi-p encountered by most of the birders there that day. In eighteen short months, I had come from being a novice birdwatcher to discussing the nuances of the covert feather patterns of Semi-ps versus Little Stints with Ireland's top birders.

That day I also saw other North American waders, including a Baird's Sandpiper and four Buff-breasted Sandpipers. The latter is an exquisite bird; with its large black eyes, beautifully patterned upperparts and sandy-buff underparts, it is possibly the most gorgeous shorebird in the world. It has a special quality that few other birds possess and which is hard to define. It is delicate, yet robust. It has a subtle beauty. You need to have the Buff-b experience to understand what I mean.

Buff-breasted Sandpipers are also incredibly tame, as they breed in the high Arctic tundra of north-western Canada and, as a result, rarely encounter humans. They know no fear of humans and simply walk by you as if you don't exist. Buff-breasted Sandpipers are among my top five birds in the world.

This September I was going back to Cape Clear. I was buzzing with excitement as I boarded the *Naomh Ciarán* in Baltimore on 27 September. I was staying in the Bird Obs and had a full two weeks of superb autumn birding ahead of me. This time, I would be arriving on Cape Clear as a fully fledged birder. I knew the lay of the island and where birds might be found. I knew how to 'do' the island. There were lots of birders already on the island and, with talk of good numbers of migrants around, there was a great air of expectancy.

Among the birders on the island were a few guys from Northern Ireland, including Anthony McGeehan from Belfast. It was the first time I had really met this young northern birder, and over the following week I discovered that Anthony was perhaps the sharpest birder I had ever met. His knowledge was in a different league from most and his sharpness of eye was second to none. Since then, I consider any time spent with Anthony in the field to be a master class in the art of all things bird.

My two weeks on Cape Clear were superb. Without going into all of the daily details, I will simply say that I 'had' lots of new birds, including Icterine Warbler, Dotterel, Red-breasted Flycatcher, Richard's Pipit and Scarlet Grosbeak (now known as Common Rosefinch). I even saw my first Lesser Whitethroat. When I left Cape Clear Island on 11 October 1980, my Irish list stood at 198. I had seen 37 new species since I had last been on the island. My target of 200 by the end of the year was now within my grasp.

On 2 November, my Irish list crept up to 199. We were birding the coastal stretch from Greystones to Kilcoole in Co. Wicklow when we found a Grey Phalarope swimming in the sea. These diminutive waders are known for the fact that they are strong swimmers and are often seen just bobbing on the top of the water in storm-force winds and huge waves. I could now almost taste and touch my 200th Irish tick. But what bird would it be?

On Thursday, 13 November, I got a phone call. It was amazing news. A wheatear had been found at Knockadoon Head in Co. Cork the previous weekend. It was a Pied Wheatear, a bird of south-eastern Europe and the Middle East. It was the first recorded sighting in Ireland. It's hard to believe in this modern era of instant news, but in 1980 news of the Pied Wheatear was sent to Dublin by letter. So, even though the bird had been first seen five days earlier, news was really only breaking in Dublin on 12 and 13 November.

Muller, Dowdall, Fitzer, Gracer and many others went to Co. Cork the following day. The news was positive: the bird was still present and showing well. Ronan could not go until 15 November so I had to wait. This could be my 200th Irish tick and I could not have asked for a better bird to mark the occasion. I read up on all I needed to know about Pied Wheatear. Other possible 'eastern' wheatears can be confused with Pied. This bird was a young male, very dark on the upperparts with a dark throat and creamy underparts. When it flew, the extensive white rump, which extends onto the back, would be very obvious, as would the striking black and white tail. I had all its details memorised.

That night, I walked out into the back garden at home and looked up. It was a starry night.

'That's a very clear night,' Da said, as he joined me outside. Ma and Da knew that this bird was a big moment in my birding life. 'Let's hope it's lashing rain in Cork, eh?' he added.

He knew that, before a twitch, I dreaded starry nights. The bird had to leave some time and such weather could inspire this wonderful bird to move on, and I needed to see it. Everything I had worked for during the year would be worth it if I did.

I did not sleep a wink that night.

It was 3 a.m. when we set off for Co. Cork. I was delighted to discover that, overnight, a rain front had indeed arrived at the southern coast and was moving northwards across the country. Driving conditions were not the best, but at least it meant the wheatear was less likely to have left. As we arrived at the outskirts of Cork, the weather cleared and the first rays of early-morning sunshine broke through the clouds. We continued along the narrow country roads, making our way to Knockadoon Head. The lads who had seen the bird the day before had told us that it favoured the area near the pier and the small stony beach close by. We would begin our search here.

As we drove towards the harbour area of Knockadoon, my heart was beating and my stomach was in a twist. I was almost afraid to be positive. What if the bird had left or had been taken by a Sparrowhawk? We pulled up alongside a pile of lobster pots on the small pier. Ronan turned off the engine and suddenly let out a loud 'Jesus Christ!'

I looked at him.

'There it is!' he said, looking at me.

'Where?'

'Right there behind you!'

I looked around slowly and there, perched on a lobster pot no more than an arm's length away, was the Pied Wheatear. I could not believe my eyes. My 200th Irish bird was there in front of me. In birding, there are moments that are etched for ever into your consciousness – this was one of those moments. It is an image that will stay with me for as long as I live (or until dementia sets in). It was a superb bird. It was a bird to savour. We watched it from the car without bins for a few minutes before it flew off along the pier and onto the beach, showing its extensive white rump and striking black and white tail – those distinctive details I had memorised.

We fell out of the car, exhausted after the six-hour drive. However, we were also ecstatic. I punched the air in celebration and there were handshakes all round. It was 9 a.m. and there were no other birders around. We got our scopes and bins, and walked down to the beach where the Pied Wheatear was happily feeding on insects in among the washed-up seaweed. Another bird flew up from the beach and landed on the rocks just beyond where the wheatear was feeding. It was a dull grey bird, almost Robin-like in shape. It flicked its orange-red tail – it was a Black Redstart. It was my first Black Red. In the space of a few minutes I had seen my 200th and my 201st Irish birds.

I smiled to myself. I had done it. I had achieved what I had set out to achieve: I had passed the magical 200 mark and was now one of the top listers in the country. These thoughts were interrupted by the sudden arrival of a car just beyond the pier. It screeched to a halt and a man jumped out. He had bins around his neck and he was looking around him. I left the beach and went up to him. He saw me.

'Any sign of the Pied Ear?' he asked in an English accent.

I was only too pleased to bring him down and show him the bird. He was delighted and very obviously relieved to see it.

'This is my 200th Irish tick,' I said excitedly.

He congratulated me and took my hand. 'Well done ... eh ... ?'

'Eric ... Eric Dempsey,' I said.

We shook hands.

'Ron Johns,' he said. 'Delighted to meet you.'

I was gobsmacked. Ron Johns was *the* most renowned British twitcher of all time. He had the highest list of anyone, and here I was boasting about seeing my 200th Irish bird.

Ron was in Co. Cork because British birders counted what they saw in Ireland as part of their British list (most still do for a reason I have never fully understood) and he 'needed' Pied Wheatear; it was a tick for him. He had left London that morning on a flight to Cork Airport, hired a car and drove to Knockadoon. The Pied Wheatear was his 429th British and Irish tick. After we had watched the bird for a long time, we headed off to Ballycotton, where Ron joined us for a while. He then drove back to Cork Airport, handed back his hire car and flew back to London.

This was the first time I had ever encountered twitching of that level, but in years to come Ron's arrival at Knockadoon Head would appear very tame indeed.

Chapter 14

ALL FOR THE LOVE OF A DESERT

Being a northside Dub, I cut my birding teeth on the wide-open estuaries of Dublin Bay and the North Bull Island. There is nothing I like better than to sit at an estuary and look through thousands of gulls or waders. When I consider the main rare birds I have found in my birding life, they mostly include gulls, terns and waders. I am not a natural 'passie basher'. Let me explain: passie bashing is the term used by birders who spend their hours looking for the likes of warblers and finches on headlands in autumn. Such birds are scientifically classified as passerines (perching birds), so searching for them along hedgerows or in gullies on remote headlands or islands is called 'passie bashing'. No, give me a big estuary with 10,000 Dunlin to look through and I'm a happy man. I feel the same draw to open spaces when birding abroad.

Earning money afforded me the luxury of heading off on birding holidays abroad. Seeing new bird species, new bird families, and new countries and habitats was a wonderful experience for a young birder. I remember the excitement of entering into the darkness of my first rainforest. The vegetation was impenetrable. There were birds calling everywhere, but I just couldn't see any of them.

Rainforest birding is very frustrating. When a guide finds a bird, it is almost impossible to see it. Rainforests can also be remarkably quiet for hours, with hardly a bird to be seen. Then, within minutes, you are surrounded by a moving mixed flock of perhaps 500 birds. Your guide shouts out names and frantically points to this tree or the next as you try desperately to see the birds he is trying to show you. Then, as quickly as they appeared, the whole flock is consumed by the forest. All that's left are birders trying to piece together what birds they might have seen.

As a birder, I am definitely at my best in open areas. Of all these types of foreign habitats, my favourite are deserts. I love deserts. Having said that, my fair skin and my tendency to burn in two seconds does not exactly lend itself to being in deserts. Within a day, I am like a lobster, even with sun block of the highest factor and a hat like a flying saucer. However, that does not deter me. I recognise that I am not made to be in a desert, unlike the birds that are found in them. Being in some of the hottest, parched environments on earth and seeing species so perfectly adapted to their environment is awe-inspiring. Take a small dove-sized bird like a sandgrouse, for example. They nest out on the ground in deserts. Each morning they flock to small watering holes to drink. They then sit in the water, using their breast and belly feathers to soak up water before flying off into the desert again. They return to their nests and their chicks drink the water that has been stored in their feathers. It's like taking a drink from a soaking sponge. How wonderful is that?

At this stage of my life, I am lucky enough to have visited many deserts in the Middle East, Asia, Australia and North America. I have also fulfilled a lifetime ambition to see the enormous sand dunes of the Namib Desert in southern Africa. I remember walking in over 40-degree heat to see the endemic Dune Lark amid the rich red sands of the Namib. The heat of the sand burned my feet through the thick soles of my hiking boots. When I eventually found a small flock of larks, my bins were almost too hot to handle. But the birds seemed totally oblivious to the scorching heat. Having watched them for a few minutes, I beat a hasty retreat to the comfort of an air-conditioned jeep. There is no doubting that all desert species (birds, reptiles, insects and mammals) are all highly evolved to cope with such conditions. Perhaps that is why I admire them so much.

My first desert experience was none other than the great Sahara, which I visited on a trip to Tunisia in North Africa in September 1983. The country had only recently opened new areas to tourism and so I jumped at the opportunity to do some birding there. It would be my first time on the African continent and that alone was a big draw. I took a package holiday and headed off on a plane full of sun worshippers, filled with excitement and anticipation. I was a 22-year-old birder and keen to see what I could see. The package based me at two different holiday resorts. Unfortunately, car hire was not an option at that time so I was happy to simply go birding each day in the local lagoons and scrubland. This was good enough for the first week, but by the time the second week arrived I was really looking to go further afield. So, when a chance came to take a three-day bus tour into the desert, I didn't think twice.

Let me immediately state that organised bus tours and birding generally do not blend well. However, as a young birder, I was more than happy to tag along on this tour. While the others were sightseeing, I went walking and birding. It was the perfect solution and it worked for me. The desert was spectacular. I was seeing Long-legged Buzzards, Mourning and Black Wheatears, and Fulvous Babblers. I watched Swallows feed around the small villages before heading out across the vastness of the sand that stretched to the horizon. Watching them disappear into the Sahara as they began one of the toughest legs of their autumn migrations is a memory that is still etched clearly on my mind. I was in desert birding heaven.

The bus trip took us for two nights to an old French Foreign Legion fort. It was basic to say the least. We arrived on the first night in darkness and many of the group gladly jumped into the newly built pool to cool off after the long day. I think many regretted that the following morning when the filth of the pool was revealed to them in daylight. The water was brown and a scum of grease had formed along the sides. Drowned insects floated around like freshly blown autumn leaves. To say that this place was not exactly clean is to put it mildly.

I had not gone swimming the night before, choosing instead to try and get some sleep. We had eaten in a small, dusty restaurant in an area called Matmata. This is a very ancient Berber settlement, known for the fact that the tribes that settled there lived in dug-out caves. These settlements were not known to outsiders until 1967. With the opening up of tourism in Tunisia, Matmata became a popular tourist stop-off point. The restaurant we had eaten in was actually in a cave and was dark. We were served a variety of food in various sauces, which we ate quite happily. For the record, I was a notoriously fussy eater (I still am, but to a much lesser extent). As a result, I carefully picked at this selection of meats and couscous. The sauces were quite strong and disguised the taste of the meat. As I ate, I looked at the bones that were left on my plate. They were small and I jokingly said that I was sure we were eating rats. Everyone laughed.

That night, as everyone else swam in the disgusting pool, I didn't feel great. My stomach was telling me that no food would have been better than that food. By morning, I was in real trouble with cramps and a severe dose of 'holiday tummy', as some like to politely call it. I stayed in my room and visited the hole in the ground (the communal toilet on the corridor) many times during the morning. I obviously had a bit of food poisoning. Thankfully I had some tablets, which eased the symptoms.

Birding was curtailed. In fact, everything was curtailed. I dosed myself up with more tablets and that was just about sufficient to get me back to the comfort of the hotel room in the main tourist resort. The bus journey to the resort was sheer hell and I had nausea, cramps and bouts of diarrhoea. I sat in my hotel room, thankful for the luxury of my own toilet. I drank water and ate very little. My cramps continued for the entire following day. I went to a local pharmacist who gave me some powders to dissolve and drink. They eased my troubles to a small extent, but I was still feeling dreadful.

Over the last two days of the holiday, I hardly ventured outdoors. I was feeling weak and I could not risk going far from my room in case I was taken off guard. By the end of the week I was more than happy to be leaving Tunisia. In fact, I couldn't get on board the flight back home quick enough and, when I did eventually touch down in Dublin, I almost got on my knees and kissed the ground. But I was too ill to risk even that.

Anyone who has travelled knows exactly what it's like. Even when you visit modern, clean destinations, travelling and holiday tummy can often go hand-in-hand. But generally, the further you go off the beaten track, the greater your chances of suffering a bout of illness. After eating in a cave in the Sahara Desert, it was not too surprising that 'delicate Eric' got a dose. Getting home and back to more familiar food normally does the trick. Usually you are fine within days and the experience of holiday tummy is overtaken by the great memories of the holiday you've just had. This was what I expected. I returned home on Saturday evening and rested all day Sunday. The symptoms continued all weekend.

By Monday morning, I was worse than ever. I could not go to work. I was still living at home and was more than happy to be looked after by Ma, who bought all sorts of drugs and powders for me to drink. By the end of the week, I convinced everyone around me that I was well enough to go back to work. My main worry was that, if I didn't go back to work, I would fail to qualify for the week's cultural leave from work that had been awarded to me.

Cultural leave was unpaid leave granted to civil servants who were making a contribution to the culture of a Gaeltacht area. It was usually granted to Irish-speakers who went off to Connemara or Co. Kerry to do courses on the Irish language or to teach music. Somehow (and, to this day, I don't know how), I managed to convince those in power that I was making a significant contribution to the culture of Cape Clear Island (a Gaeltacht region) by staying in the Bird Observatory. I was chancing my arm but, to my utter amazement, they granted me a week of cultural leave. I planned on visiting Cape Clear Island during that week and nothing was going to stop me. There was no doubting it: any birds I'd see that week would be a superb contribution to the culture of Cape Clear.

So, while I still did not feel 100 per cent, I felt I had recovered sufficiently to go birding again.

Chapter 15 ✦

A SHORT-TOED LARK AND A COMMON SANDPIPER BETWEEN FRIENDS

On Saturday, 1 October 1983, I once again found myself arriving on Cape Clear for a short, week-long birding trip. I didn't have much money after Tunisia, but this didn't bother me. I had a full bonus week off work to spend at the best birding location in Ireland and at the best time of the year.

Arriving fresh on Cape Clear in the autumn, you are always filled with a sense of excitement and anticipation. Cape entices you back with a promise of new birds and new birding experiences. You step on board the *Naomh Ciarán* and, in the hour it takes to sail from Baltimore to Cape Clear, it is as if you are transported to another world, a world away from the pressures of normal life. As the boat swings into North Harbour and you catch a glimpse of Cotters Garden, you know you've arrived in the magical world of birding.

It was a foggy day with light winds when I stepped off the boat and onto the pier. I love foggy days on Cape. There is a mysterious air to places like the West Bog, from where you could hear the booming foghorn from Fastnet Rock lighthouse through the gloom. Foggy conditions are also good for grounding migrants, which are reluctant to fly long distances in poor visibility.

This was my fourth autumn visit in a row to Cape. At the Bird Obs, I said hello to Tim Collins, who was acting warden at the time, found out which room I was staying in and quickly set off into the misty shroud that hung over the island. For the moment, there were only a handful of birders present, which offered a great opportunity to find a good bird before the main posse arrived. From tomorrow, the Obs would be full for the week.

As always, Cotters Garden was my first port of call. It was a good place for getting a feel for what might be around based on what birds were flitting near the trees. A few Chiffchaffs and a Pied Flycatcher were good omens. I spent about twenty minutes watching the garden before moving on. I walked up Cotters Hill. As I reached the top, I pondered what direction I might take. Would I go towards the Waist and then on to the youth hostel and post office? Or would I go along the Low Road to East Bog and Wheatear Field? Then again, maybe the High Road, towards Lough Errul and West Bog would be a better option?

As I reflected on which option might be the most bird-productive, I saw a young scrawny guy who I didn't recognise come running down the High Road. He had a pair of ancient bins swinging wildly around his neck. He looked about seventeen or eighteen years old, and he was wearing a dirty green rain jacket, faded jeans and a pair of black wellington boots. He had scraggy hair and he looked a little dishevelled. Seeing that I was a birder, he ran up to me, stopping abruptly in front of me. His glasses, which were perched at the tip of his nose, were fogged up with the misty rain that was falling lightly. He pushed them back up his nose.

'There's a Short-toed Lark at Lough Errul,' he announced excitedly in a strong northern accent.

'Thanks,' I answered nonchalantly. 'I might head up that direction later on. Thanks for letting me know.'

He gave me a puzzled look. I think he was taken aback by my lack of interest. Then he ran off down towards the Obs.

I watched him go off down the hill. Twice I had experienced Short-toed Larks being 'called' on Cape Clear by birders. One had turned out to be a Skylark and the other, remarkably, a young Linnet. 'Young birders come here and expect that every bird they see will be a rare one,' I found myself thinking as I watched him reach the end of Cotters Hill and run past the Club. At the age of twenty-two, I was now a well-seasoned Cape Clear birder.

Still, his intervention influenced me into a decision to head up the High Road, look for his supposed Short-toed Lark and then head towards West Bog. I reached Lough Errul and looked along the short grassy bank on its north shore, where I had seen a Short-toed Lark in 1979. There was a Yellow Wagtail feeding with some White Wagtails, just like there had been in 1979. I walked along the edge of the lake and stood enjoying the Yellow Wag. There were Linnets around too. Then, along the top end of the bank, I saw a lark. I raised my bins – it was a Skylark. I rolled my eyes to heaven. Just what I thought: this young guy sees a Skylark, and, because he's on Cape Clear, it suddenly becomes a rarity, a Short-toed Lark.

I waited another while. There was nothing else of note except a Common Sandpiper at the far end of the lake. I was about to turn to leave when I saw a bird move out from the longer grass. It walked (shuffled might be a better description) out into the open. It looked sandy in colour. It looked clean. It looked dumpy. It looked like a bloody Short-toed Lark. I raised my bins and there was a pristine Short-toed Lark. He was right.

I set my scope up and started watching this beauty of a bird. I was so concentrated on watching it that I hardly noticed this young guy arrive back. He was out of breath, but he was beaming. His face was full of relief that someone else was watching his bird. I felt a deep sense of guilt at not believing him. I felt sick that, in just four short years, I had become the kind of cocky birder that had intimidated me when I was starting off. I did not like the fact that I had been acting the know-it-all and the doubting Thomas. I felt ashamed.

'It's still here?' he asked. 'That's great.'

'Yes, great find,' I said with enthusiasm, trying to make up for my earlier lack of interest (or lack of belief).

I put out my hand and he shook it.

'I'm Eric Dempsey,' I said. 'Pleased to meet you.'

'I'm Michael O'Clery,' he replied. 'Nice to meet you. I heard you were coming onto Cape today but I wasn't sure if that was you when I met you.'

We watched the bird together for another ten minutes before the local horse arrived on the scene. I had spotted an enormous mushroom just behind where the lark was feeding and had decided it would be great for dinner. Michael had also spotted it and we said that, as it was so big, we would share it. Wild mushrooms grew all over the island and were a wonderful addition to boring dinners. However, before we could pick the mushroom, the horse duly took one step towards us, eyed up 'our mushroom' and, in one bite, consumed the whole thing. Needless to say, dinner that night did not include that mushroom. We didn't know it at the time, but a young German hiker, who had arrived on the same boat as me that morning and who was staying in the youth hostel, had the same idea of picking mushrooms for dinner that evening. He was last seen walking along the cliffs on the old lighthouse side of the island. He didn't return that evening. We all assisted in the search for him over the following days. His bag of mushrooms was found but he was never seen again. It was a stark reminder of the danger of the cliffs around the island.

As other birders gathered to watch the lark, I decided to head off towards West Bog. Michael (from here on, I may refer to him as Mick, as I always do) joined me. We saw precious little else bar a Whinchat, but I showed him places where I had seen some good birds during previous visits.

Later on at the Obs, Mick wrote up his description and drew several lovely pencil sketches in the Rarity Form for the log. He was a talented artist. In the pub later on, I bought him a pint in the true traditional manner. We sat talking bird with the gang, but we also discussed bird artists and drawing techniques. Arrogantly, I offered him some advice on the advantages of sketching in the field, which he politely accepted. In the Club that night, all that mattered was that we had a Short-toed Lark to celebrate.

Over the following days we headed off birding together. I quickly learned that Mick was a sharp birder with a lot of knowledge but was also humble enough to learn from others. I found that he was an honest birder, an honesty that prevails in his birding to this day. The week produced some major finds, including a very tame Dotterel in the Wheatear Field as well as two Icterine Warblers and a beautiful Yellow-browed Warbler (these little Siberian gems are always beautiful). Mick was having the same kind of birding week that I had in 1979, with lots of new birds for him. I, on the other hand, felt like a seasoned birder, since none of these species were new for me. Not getting a tick with so many good birds around reflected my experience and vintage. That felt good in its own way.

It was great birding but my final day might not exactly be remembered for my contribution to the culture of Cape Clear. Instead it was a day when Mick and I found ourselves embroiled in a major controversy, one that was testing and challenging. As always, it was about a bird.

We had been birding the West Bog area and we were returning back via Lough Errul. We began to walk around the southern edge of the lake. It was very quiet apart from the Common Sandpiper that had been present all week.

'Hello, Eric Boy. Any good birds around?' Mary Mac greeted me with her usual warmth.

'There's nothing around at all today, Mary,' I answered. 'It's very quiet.'

She wished us luck and I said my goodbyes, as I was leaving the following morning.

As we continued along the lake edge, we flushed the sandpiper, which flew silently up along the edge of the lake before landing in the corner of the lake. That struck me as odd, because a Common Sand would usually call. The bird was sitting on a rock near the small wall that ran into the water at the west end of the lake, about a hundred metres from us. I casually turned my scope on the bird and focused in.

It was a young bird, as told by the strong barring on the wings. It was half facing me and I thought the legs looked quite bright. It flew off the rock but landed on the far side of the wall, out of view. I had small alarm bells going off in my head. Those legs looked a little too bright and the wing barring too strong for Common Sand. I told Mick that I had some suspicions that this bird might be a Spotted Sandpiper, a very rare North American vagrant (their version of Common Sand). We approached the wall carefully and peered over it. The bird had moved along the shore and was too busy feeding to be overly concerned by our presence.

I focused in again. The tail looked quite short – a good feature for Spotted. The bill looked two-toned – another good feature. The breast sides looked smudgy ... This was looking good. The edges of the tertials (the long wing feathers that cover the primary and secondary feathers) looked plain and lacked the usual notched edges seen on Common Sandpiper. Then the bird flew off to the opposite side of the lake again.

I looked at Mick.

'That's a feckin' Spotted Sand!' I announced.

Mick was unfamiliar with the species, but when we went through what I remembered as being the key identification features of Spotted versus Common, he agreed that he too had seen all of the main features. I had seen a Spotted Sand at Clonakilty in Co. Cork the previous year and this bird appeared to be exactly like that one. This had to be a Spotted Sand. We spent another hour or two trying to get a better view, but the bird always remained at too great a distance for its more subtle features to be observed. However, even at a distance, the tail looked far too short for Common Sand.

We returned to the Obs to inform the others about what we had seen. By now, the Obs was full and, with no fresh bird arrivals, everyone was down in the house either cooking or reading while they waited their turn at the cooker. We announced what we had seen and informed them that we thought the Common Sand present all week was in fact a Spotted Sandpiper. There was silence before someone spoke.

'You're a fuckin' stringer, Dempsey. That's no Spotted Sand! I've watched that bird loads of times this week. It's a fuckin' Common Sand. You're a right fuckin' stringer, Dempsey!'

This verbal assault took us by surprise. Of course, what we were doing was challenging the skills of other birders who had possibly seen the bird well many times during the week. That was not what we set out to do, but that was obviously how it was taken.

I didn't take their bait. Instead, I went through all of the features we had observed and told them why we thought it was a Spotted. Poor Mick. This was possibly the first time he had encountered this level of bird-identification argument. I was confident of what I had seen. I had the experience of the Clonakilty bird to draw on. Mick was depending on what he had observed but had no experience to back up his views. His opinion was very quickly dismissed.

However, I stood my ground. Those in the Obs were divided. People either opted out of the argument, saying they hadn't seen the bird well enough, or rowed in with those who felt it was just a Common Sand. The argument became heated. Mick stood firmly behind our view that the bird was a Spotted Sand. It was Mick and me against five very vocal and experienced Dublin birders.

In the Club that evening, the voracious slagging continued unabated. We were called 'stringers', amongst many other things. To be called a stringer is the ultimate insult in birding. There are unwritten rules between birders. There is a sacred trust. It would be easy to claim you have seen a rare bird, tell everyone about it and then announce it has flown away before anyone else can see it. This does happen: birds do fly away and not all birds found will stay long enough for others to see it. When this happens you have to take the person's word for it. You need to trust that they have indeed seen the bird and are not 'stringing you along'. In most cases, this trust is a given. However, stringers are birders who frequently claim to see rare birds that others somehow never see, or they claim that a common bird they've seen is in fact a much rarer species. Being labelled a stringer is the worst possible thing for a birder. Once labelled, that birder, regardless of what they achieve in life, will go to their grave as a stringer; and their descendants for the next fifty generations will always be under suspicion. You could be the man or woman who cures cancer but you would always be referred to within the birding tribe as 'the stringer who cured cancer'. Stringers break the sacred trust among birders and that is never forgotten. So, to be called a stringer was a very serious charge against me.

That night, I didn't sleep well. I was heading off the island the following morning, but I was determined to prove that I was no stringer. I got up at dawn and raced up to the lake to see if I could find the bird. There was no sign of it. I just about made it back to the North Harbour in time to catch the morning ferry off the island.

I so wished that I had found that sandpiper at the lake before I left. I hoped that it would be re-found and that it hadn't flown off that night, never to be seen again. I was broke by the time I left Cape Clear and only had a tenner that Mick had loaned me to get me home. I would catch the bus from Baltimore to Cork but then try to hitch to Dublin. It was going to be a long day but my mind was filled with the fury of the events of the night before. I was strolling up the hill from Baltimore Pier deep in thought when I almost bumped into two people. It was my old Cape Clear friends, Tony Lancaster and Pat Hamilton. They were walking down to the pier to get the early sailing to the island.

I barely said hello to them before filling them in on the sandpiper on the lake and the trouble it had landed me in. As they were getting on board, I told them of the features I had seen. I also let them know that there was a great young birder called Michael O'Clery on the island. He'd show them the sketches he had done of the bird.

'I am sure it's a Spotted Sand,' I shouted as I waved them off. 'Please go check it out!'

The journey back to Dublin was a long series of short lifts from one village to another. Eventually I reached home late in the night.

I was exhausted by the effort of hitching home. I was exhausted by the mental strain of the argument that had taken place on Cape Clear about the sandpiper. I was emotionally exhausted by the fear that if the bird was re-found and proved to be just a Common Sand, then my reputation would be in ruins. Mick's would remain intact, as birders would take the view that, as a young birder, he had been badly influenced by me. I was filled with self-doubt and anger.

Some days later I got a phone call from Mick. He was still on Cape Clear. The line was faint and fuzzy (as it always was when you called from the phonebox at the island post office), but through the noise I could make out the magical words:

'It *is* a Spotted Sand, Eric. We were right!'

The write-up that followed in the 'Cape Clear Bird Observatory Report Number Eighteen, 1983–1984' read quite simply:

Spotted Sandpiper
1st–15th October 1983   5th Irish Record.

On 1st October 1983, a wader was located at L Errul and identified as a Common Sandpiper *Actitis hypolecos*. The bird was extremely wary and it was virtually impossible to obtain good views. On 7th October E. Dempsey and M. O'Clery saw the bird reasonably well and put forward the possibility that it could be a Spotted Sandpiper *A. macularia* …

That was the first time the names Michael O'Clery and Eric Dempsey appeared in print together. Neither of us knew it back then, but we would go on to become life-long friends, have many birding adventures together, become part of each other's families and travel together to some of the best birding destinations in the world. Little did we realise that, just ten years later, almost to the day on which we had first met on Cape Clear Island, our first book, *The Complete Guide to Ireland's Birds*, would be published.

Chapter 16 ⇥

THE MINISTER FOR HEALTH

By the end of October 1983, the symptoms of the illness I had contracted in Tunisia returned with a new vengeance. I remember the moment well. I was on my way into work on the bus and suddenly felt cramps. My stomach was churning. I broke into a cold sweat. I looked around the full morning rush-hour bus that was slowly weaving its way into the city centre through heavy traffic. I held on. At my stop, I pushed my way off the bus and ran as fast as I could to my office at the GPO. I ran up the stairs and along the corridor to the toilets … but it was too late. I didn't make it. The embarrassment was overwhelming. I tried to change and clean my clothes as best as I could before sneaking out of the toilet. I went outside, got into a taxi and went home. I was almost in tears. What was going on?

That day I visited my local GP and he suggested that I just had a bit of food poisoning. It was nothing to worry about.

Over the following weeks, things did not improve. In fact, the bouts of diarrhoea worsened. It seemed that every time I ate something, it would only stay in my body for less than an hour. I sweated every time I stepped on board a bus to go to work. I had figured out potential places along the route that I could run to should I get caught out. My birding was seriously affected, although I carried enough tablets to get me through long days in the field. Worse than that, my energy levels began to reach new lows.

Months passed and, whilst I fondly remember the early weeks of 1984 for the beauty of my first adult Ross's Gull, they are also remembered for my continuing illness. By now I was beginning to lose weight. Visits to my GP consisted of him attempting to guess what might be causing my symptoms.

'Worms!' he announced one day and so he began treating me for worms. I felt like a cat at the vet.

Like everything else, my worm treatment did not work. I was ill every single day. It became so bad that, at work, I was given the key to the 'EO's toilet'. Executive Officers were the mangers of us lowly Clerical Assistants and they had their own, exclusive toilet. It was close to my office, and so, through the understanding and effort of my boss, a key was secured.

As 1984 moved on, there was no sign of my illness improving. Among the birding tribe, it was obvious that I was ill. Weight was falling off me. I was pale and had dark circles around my eyes. I became known as 'The Minister for Health' by some of the birders. It was funny but few knew how sick I was. In fact, I didn't know just how ill I was or what was wrong with me. By the time I turned twenty-three, in July 1984, I had lost almost two stone in weight.

My GP made more suggestions as to what my problem was. They ranged from a parasite to food allergies. However, at no point did he think it was anything to worry about. He did, however, refer me to a consultant. I was a long way down the waiting list and it would be a long time before he could see me.

As 1984 came to an end and the new year commenced, I was still very ill and losing weight each day. I was missing days at work. I was having 'accidents' and I was swallowing far too many tablets in my efforts to prevent such accidents. My confidence was shattered. How could I face sitting on a bus with the fear of humiliation a constant reality? How could I face my work colleagues, young lads and girls, if I was caught out in the office? I began to fear venturing outside. I was becoming deeply depressed.

By now, poor Ma and Da were so worried that they decided something had to be done. In the Mater Hospital, where Da worked, it was not the accepted custom to approach one of the consultants to seek a favour. However, he did get on well with one of the sisters who was a senior manager in the hospital. She listened carefully to what Da told her and was horrified by what she heard. She arranged for me to be given an appointment to meet the senior gastrointestinal consultant.

Ma was delighted that at last I would get to see someone who might help, and so was I. But, in truth, I was also very worried at what he might find. Sometimes the truth is not what you want to hear. So, with a sense of both relief and fear, I went in to meet the consultant. At this stage, I was skeletal and weak, and weighed less than seven stone.

I waited my turn to be called and then was taken in to meet a tall young man. He had a gentle, polite manner and he introduced himself as Dr Crowe. I smiled. This had to be a good omen. He did not smile when he saw me. He was shocked at my condition.

We sat down and he took lengthy notes about my medical history. He was horrified to learn that I had been ill for almost eighteen months. He was horrified to discover my weight, which was a full three or four stone below my weight before I went to Tunisia. I sensed that he was worried by what he might discover. I was with him for over an hour. He took blood samples, asked me to provide various other samples, and weighed and measured me. Before I left, he assured me that he would do his utmost to discover what was wrong with me.

I left the Mater and walked back to work. I had barely sat down at my desk in my office when I was called to take a phone call. It was the Mater Hospital.

'Dr Crowe wants you to report to the Mater Hospital for immediate treatment,' the voice told me.

'When should I report?' I asked.

'Right now!' came the unexpected reply. 'He wants you in a ward by this afternoon.'

I was in a state of shock. I really didn't expect this. It also frightened me, because I felt that this must be bad. Why else would he want to get me into hospital immediately? I rang home to let Ma know what was going on. I hadn't even had time to ring her to let her know how the appointment had gone, and now I was asking her to put a bag together for me and bring it to the hospital. I explained the situation to my boss and left the office. I would not return to work again for over six months.

Over the following days, I underwent a whole series of incredibly invasive tests in hospital. Tubes were shoved into various parts of my body. I was examined from top to bottom (excuse the pun) by a wide range of doctors and nurses. My stools became the centre of attention. I was a young guy and the embarrassment of having young nurses and old sisters shoving tubes up inside me was hard to take. However, I knew that this was for my own good and that, if I wanted to get well again, I had to put up with it. Dr Crowe also performed a colonoscopy on me, where a small camera was inserted and pushed along my entire bowel to see what things were like inside me. This would reveal a lot.

Following three days of tests and examinations, Dr Crowe called to see me in the ward. He spoke quietly and explained everything in detail. It was a serious situation. My entire bowel and colon were incredibly ulcerated. Things were so bad that he felt there might be no option but to remove them and insert a colostomy bag. I was devastated. This was an irreversible treatment in 1985. All sorts of thoughts raced through my mind. I would spend the rest of my life with a bag of crap that had to emptied and changed. How could I go birding? How could I travel on a plane again? I had read that this was a major problem due to air pressure. Would that mean I would never see some of those wonderful exotic birds I so longed to see? Would I ever meet a girl who would want to go out with a fella who had a bag of crap attached to him? My world as I knew it was coming to an end.

THE MINISTER FOR HEALTH

Dr Crowe recognised what this meant for me, but his following words brought things back into sharp focus.

'To be honest with you, Eric, I am relieved that this is all we found. I feared the worst when we started our examinations.'

His words hit me like a thunderbolt. Yes, having a colostomy bag was bad, but the alternative was worse. It was a sobering thought. I could have been facing my own mortality right at this moment. As a young man, thoughts of my mortality had never really entered my mind. A bag was an inconvenience but not a death sentence. At least I would be well enough to go out birding, albeit in a more restricted way.

'However,' Dr Crowe continued, 'before we do that, I am going to try to heal the walls of your bowels and colon. Let's see if it works, eh?'

I felt massive relief. It was not a *fait accompli*: there was a chance to reverse the damage. So, with great enthusiasm, I accepted everything they threw at me. The ordeal of learning how to do retention enemas still stays with me. This involved inserting long tubes into my rear, through which steroids were pumped into my bowel. I was required to learn how to do this myself. Once the steroids were administered, I had to remain almost upside down for over two hours to help the contents of the enema to flow into my system. More than once, I had messy accidents, but I had to get on with it. I was also taking drugs of all shapes and sizes, and drinking two litres of supplement drinks a day. These were nasty, thick milk-like drinks. My stomach still retches at the thought of them. But I gladly swilled them down like they were freshly pulled pints of Guinness. I so desperately wanted to get well again. I wanted to feel the breeze on my face. I wanted to feel bins around my neck again.

After three weeks of this intensive treatment, I began to feel better. The daily stool examination reported improvements. I went for a second colonoscopy. This was the make or break moment. If the treatments showed no sign of improving my condition inside, then 'the bag' was the next step. I didn't sleep well the night before that examination. I feared the worst, but in a strange way I'd also accepted that if it was to be, then it was to be.

I attended Dr Crowe's clinic the following morning and was given my pre-examination treatment: an enema that, unlike the retention enema, was designed to clear my system. I called this one the 'explosive enema', and anyone who has ever had one will understand why. Once my system was cleared, I was brought in for my examination. The colonoscopy procedure was a little painful and was always done without anaesthetic. The nurses were superb, encouraging and professional. I could not see what was going on, but I listened intently to gauge Dr Crowe's reaction.

It was all very silent. I waited.

'Fantastic!'

This was the only word I recall him saying. He said more but I had tears of relief streaming down my face. The nurse smiled and gave my hand a reassuring squeeze. The examination continued without another word being spoken. When it was over, I sat up and turned around to see a smiling Dr Crowe.

'The only bag you'll be carrying will be your camera bag!' he announced.

I could not find the words to thank him enough. I still thank him to this day.

After a total of five weeks in hospital, I returned home to continue my treatment there. Poor Ma and Da had been worried sick and my occasional accidents were dealt with with smiles and no fuss.

I remember well the first day I was well enough to go out birding. I went to Swords and saw a superb male Ring-necked Duck (the North American version of Tufted Duck). Looking through a telescope and lifting my bins never felt so good. I realised how easy it was to take these simple things in life for granted.

By the end of March I was even back twitching again, travelling to Northern Ireland to see a White-throated Sparrow at Duncrue Street in Belfast. While there, we had to duck behind bushes when a sniper opened fire on a British army patrol. The army returned fire. Bullets flew over our heads. It was a terrifying few minutes, but I felt alive.

Eighteen months after my trip to Tunisia, I was back.

Following months of treatment at home, I had gained a lot of the weight I had lost and was feeling really well. I was no longer having to administer my retention enemas, but I was still taking a lot of tablets (which I continued to take for a further five years). They never really found out what had caused the illness, but I didn't care. I had my confidence back. I was no longer fearful of any unexpected embarrassing situations. Of course, little did I realise that I was saving the best until last.

I was called to have one more colonoscopy procedure as a matter of routine. It was important to check that the treatments had worked. I reported to the Gastrointestinal Unit, met the nurses and changed into my gown. The nurse came into my small curtained cubicle and administered an 'explosive enema'.

Before she left, she pointed to a door with the word 'Reserved' on it. This was on the main corridor and was right in front of the waiting room where a queue of over twenty people were sitting.

'That's your toilet,' she said. 'When you feel the moment has arrived, just go in there.'

I was very familiar with this routine by now so I sat back and waited for the moment to arrive. And, boy, when the moment arrives, it arrives with great urgency. When I sensed the gurgling volcanic eruption gathering pace inside me, I went to 'my toilet' and opened the door. There, standing before me, was a well-dressed dark-haired lady in her forties. She was leaning in towards the mirror with her right arm curled like someone doing synchronised swimming in the Olympics as she applied copious amounts of lipstick to her puckered lips. She looked at me, indignant at being disturbed.

'Eh, this is reserved,' I explained.

She sighed at having to deal with me.

'I won't be long,' she retorted snappily.

I felt my volcanic eruption building up steam.

'I need to use this toilet,' I said with urgency.

'I told you, I won't be long.'

By now, the people who were bored in the waiting room were all watching this exchange. No doubt, it was better than staring at the four walls.

'I really do need to use this toilet!' I said sharply.

'And I told you, I won't be long!' she snapped.

With that, my world exploded. It was a volcanic eruption of epic proportions. I suspect there may have even been mini pyroclastic flows. I stood in a pool of my own liquefied excrement. It was running down my legs. It had destroyed my gown. It was lying all over the floor and on the walls of the corridor. I looked around. The people in the waiting room sat in shock with their mouths wide open. I looked down at myself. I was a mess. I then looked back at this woman. She was looking at me with wild eyes. Her mouth was still puckered and her outstretched right arm still held her lipstick. She was frozen in that posture.

'When I said I really needed to use this toilet, I wasn't joking,' I said calmly.

She said nothing, but, with her right arm still outstretched, her lipstick still in her hand and her lips still puckered, she took exaggerated giant steps over the pool of my liquefied mess and quietly slipped away. I watched her go, her arm still outstretched and her lipstick still in her hand.

I sat in the loo and cried, but they were not just tears of mortification; they were also tears of laughter as I remembered the face of the lady with the lipstick.

Almost thirty years later, I can still see her clearly. I can only imagine that over the years she has told the story of how she was 'once putting on lipstick in a hospital toilet when …'

And almost thirty years later, as I put a pair of bins around my neck to go birding or simply sit listening to birds singing in my garden in Wicklow, I once again remind myself that the simple things in life are so precious. At times it really is too easy to forget that.

Chapter 17

MY BIRDING *ANNUS HORRIBILIS*

I never thought this day would come. For those birders who have known me for many years, this will certainly come as a shock. After almost thirty years, I am ready to deal with 1985. I am about to write it all down. My therapist will be so proud of me.

You see, for a long time, if I was asked to count the years of the 1980s, I would say, '1980, 1981, 1982, 1983, 1984 … 1986, 1987, 1988 and 1989'. I had blocked the year 1985 from my birding mind.

That's not to say that, in 1985, I didn't fully appreciate the fact that I was up and about after being so ill. I was birding and I was feeling well. People were happy to see me back in the field and, like a prodigal son, I was welcomed back into the birding family. I was even offered lifts from several birders. However, while my illness might have tested me in the physical sense, what was about to unfold during the rest of 1985 would test my birding emotions to the limit.

1985 was my birding *annus horribilis*.

During April and early May, Ronan wasn't birding much, so I was mostly out and about in the local Dublin spots. This suited me well as I was still on various treatments and wasn't keen on wandering too far. However, on 18 May, with the winds light and variable and warm weather forecast for several days, I was more than delighted to avail of an offer to be a runner in a car full of top Dublin birders. We headed out to Great Saltee Island, off Co. Wexford. Such weather is ideal for the arrival of spring migrants and we weren't wrong that day.

As we landed on the island, it was obvious there had been a fall of birds, and we saw several Turtle Doves, lots of common warblers and a superb female Ring Ouzel. When weather conditions result in the arrival or grounding of lots of migrants, it is often referred to as a 'fall' – it's as if the birds are falling from the sky. As the morning wore on, birders spread out in search of birds until a shout from near 'the garden' had us all racing back to the area of the house (the only house on the island). Someone had seen a stunning male Bluethroat. This would have been a tick for most of us. It was a red-spotted Bluethroat (Northern European birds have a red spot in the centre of their blue throat, while more Central and Southern European birds show a white spot). We searched high and low for that bird but it was never seen again.

It is a dreadful feeling to leave somewhere knowing that you've missed a tick. Almost everyone in our group on board the 5.30 p.m. boat to Kilmore Quay had the same feeling. Only a handful of birders had managed to glimpse the Bluethroat. Some other birders were staying on the island for the night. I was very pensive; I had a 'you win some, you lose some' air about me.

As we neared Kilmore Quay, the local church bell sounded the Angelus. One of our birding group, our driver for the day, was a man of religion and, upon hearing the bell, dropped to his knees on the deck of the boat and began praying. The others bowed their heads and joined their hands together in respectful silence as he prayed. I think more than one of them were also saying their Hail Marys. I was respectfully quiet, but was more interested in watching a Storm Petrel that was feeding just offshore as we rounded into the quay. When the bells stopped ringing, our kneeling birder friend arose and, as if inspired by prayer, announced that we were heading to Hook Head. Since he was the driver, if he said we were going to Hook, we were going to Hook.

We arrived on Hook Head in the early evening and stopped at Lark Cottage, one of the main gardens. This usually offered a good insight as to whether there were migrants around. We climbed the locked gate and began slowly walking through the garden. A Spotted Flycatcher was sitting out in the open to the left so I went that way to get a good look at it. The other three went to the right. I was no more than ten metres away from them when I heard a shout.

'Nightjar!'

In a panic, I ran to where the lads were standing. They were beaming and bristling with excitement.

'Where is it?' I asked.

Their look told me what I needed to know. It was gone.

Apparently, as they came into a clearing, they flushed a roosting Nightjar from a small tree. It flew around them and left the garden. Nightjar was a tick for us all. They had all seen it whilst I was getting a better look at the Spotted Fly.

We spread out in the hope of re-finding the bird, but, despite extensive and exhaustive searches, we didn't see it again. I had dipped on two major birds in one day. I really did wonder if there might be a God after all. I hadn't said my prayers on the boat on the way back. In fact, I was the only one who hadn't bowed my head and joined my hands. If there was a God, maybe He or She wasn't too happy with me.

As we left Wexford, we got word that a male Bluethroat had been found that evening on Great Saltee, and it was a white-spotted bird. There had actually been two different Bluethroats on the island and we had seen neither. The pain of missing two ticks in one day is hard to describe.

It took a little while to recover from this day in Co. Wexford, but when a singing Savi's Warbler was reported from Youghal Marsh, Co. Cork, in late June, I jumped at the chance to travel down. This plain little European warbler had only been recorded once before in Ireland, so it would have been more than enough to exorcise the

Wexford demons. The bird was reported as being in the extensive reed beds, but there were only a few places that afforded good views over the area.

Lots of birders were there from dawn (us included). Birds like Savi's Warblers are known to prefer to sing at dawn and dusk, when their low-pitched song travels better in the cooler air. Sure enough, as soon as we got out of the car we could hear him singing away in the reed beds. In fact, he sang on and off all day, but we just could not see him. We remained there until dusk. He sang his little heart out, but always from deep in cover. We left in darkness without ever seeing our Savi's Warbler. This was turning into a bad run. I had now missed three ticks in a row.

Of course, what I didn't know was that this run of bad luck in the spring would fade into insignificance by the time autumn had its turn.

I won't bore you with all the sordid details of each and every day of the autumn of 1985. In fact, the real reason for not giving all of the details is that I don't think I want to torture myself too much. It is enough that I can bring myself to even mention the birds as they occurred. The autumn of 1985 was *the* autumn for North American birds. It was *the* autumn of the century, possibly of all time. The birds I did not see read like a list of who's who of mega rarities.

Short-billed Dowitcher (first Irish record) at Tacumshin Lake – missed it because I couldn't get a lift down to Co. Wexford.

I was gutted.

Red-eyed Vireo, Yellow-rumped Warbler and Indigo Bunting (first Irish record) on Cape Clear Island – missed all three because I couldn't get a lift down to Cape Clear.

I was beginning to think the great birding gods were conspiring against me.

Scarlet Tanager (two birds; second and third Irish records) at Firkeel, on the Beara Peninsula, Co. Cork – missed them both because I couldn't get a lift down to Firkeel.

Could this really be happening?

Philadelphia Vireo (first European record) and American Redstart (second Irish record) on Galley Head, Co. Cork – missed them both because I couldn't get a lift down to Galley.

I had lost my birding will to live.

As October ended, I was reduced to a gibbering mess, barely recognisable as a birder. I reflected that I had seen a few good birds during the autumn but there was a dark shadow cast by the seven major rarities that I had missed. Adding salt to the wound were the spring dips of Bluethroat, Nightjar and Savi's Warbler.

Each one of those birds is an avian superstar in its own right. To have missed one or two would have been dreadful; to have missed them all was beyond words. There was a great emptiness in my notebook, in my list and in my life.

This situation did make me realise one thing: my days as a runner were nearing an end; from now on, I needed to be in control of my own twitching. In the first week of November, I smashed my piggybank with the biggest hammer I could find, and, with some additional funding in the form a loan from Ma and Da, I bought my first car. It was a metallic-blue Golf. I now had wheels and was the master of my own birding destiny. I would not miss a good bird again. I could go wherever and whenever I chose to go. I would never allow a mega bird to slip through my fingers again.

So, when news started to break in late November that an 'interesting plover' had been seen near Blennerville outside Tralee, Co. Kerry, I was on high alert. When the bird was eventually tracked down feeding with Lapwings and identified, the news was unexpected. It was a Sociable Plover, a bird of eastern origin, and this was the first time this species had been seen in Ireland since 1908. It had arrived with a large flock of Lapwings. The news was that it 'was showing well' at Blennerville. I didn't need to think twice.

On 7 December, with my red L-plates showing clearly, I hit the road. It was my first twitch in my own car. I felt in control of my birding at last. It was a very liberating feeling.

I left early and belted south as fast as my car could take me (or as fast as I dared to drive). There were no motorways in 1985, so the road to Tralee went through various villages and towns; it was always a long journey to Co. Kerry. However, I was on a mission. This Sociable Plover would more than make up for all of the birds I had missed in October. I also loved waders, so the thoughts of this leggy, beige Lapwing-like bird really appealed to me.

Perhaps my mind was on the Sociable Plover, because I didn't see the pothole. Actually, 'pothole' is not the correct description: 'tank trap' comes closer to describing it. All I remember was the bang as I hit it. One of the front tyres exploded with the impact and the car skidded. I pressed the brake and tried to change gears, but the clutch did not work. I somehow managed to stop the car, and I assessed the damage. My front wheel was twisted, my tyre was blown off and in shreds, and my clutch had broken with the impact.

I was very thankful for the assistance of a mechanic, who took the car in and worked on it. I sat waiting all day Saturday and most of Sunday while the car was being fixed. The Sociable Plover was not just showing well, it was 'giving itself up' (a term used by birders when a bird is giving close, stunning views). I was still alive and had not crashed the car, but these thoughts, while offering some positives, really meant nothing. I only thought of the Sociable Plover. Thankfully, the mechanic got the car up and running in time for me to make a mad dash to Co. Kerry in the late afternoon.

I arrived at Blennerville with an hour of daylight left. I met birders who had seen the Sociable earlier but they told me it had flown out of the bay with the Lapwings in the afternoon. We searched every square inch of mudflats, channels and fields. There were lots of Lapwings but no sign of the Sociable.

It was the first time I ever drove home myself as a dipper. It was a very long drive and it was late by the time I reached Dublin. I was in a deep birding depression, the kind that only those who have dipped will understand, but I vowed that, if the bird was seen again, I would travel to see it. I was determined not to allow the Sociable Plover to become another statistic in my 1985 list of mega dips.

I visited Blennerville on seven more occasions, and each time I missed the bird. Then, while there in the late afternoon of 27

December 1985, I met Frank King. Frank is the grandfather of Kerry birding. We knew each other well. I had been looking for the Sociable Plover all day without success.

'Ah, Eric,' he greeted me. 'Have you seen the Sociable Plover yet?'

'No,' I replied. 'I've been out all day and checked everywhere, but no joy.'

'Well, now,' he said with a smile, 'I have good news for you.'

My heart was pounding as I listened to what he had to tell me.

'I just saw it a few minutes ago. It was flying like a butterfly with the Lapwings and has gone with the main flock onto the estuary.'

I couldn't believe it. I thanked him, hopped into the car and drove to the old windmill, from where I could see the mudflats. The year 1985 was going to end on a high after all.

I raised my bins. There were probably 10,000 Lapwings following the dropping tide. I was elated. This was the big flock and the Sociable Plover was with them. I set up the scope and started scanning through the flock. It was late and the light was fading fast. The tide was also dropping at an alarming rate and the whole flock was drifting further and further away.

Then, in amongst them, I found a pale bird. This had to be it? Surely this was it?

I couldn't see it clearly. It was in the middle of the flock. It was the right colour and size. It was hanging out with the Lapwings. This had to be the bird. If it flew, I would be able to pick it out by the beige-and-white wing pattern. Lapwings are jumpy birds. They fly if a Grey Heron flies overhead. Surely the flock would fly sooner or later.

'Where's a Peregrine when you need one?' I thought to myself as I watched this pale bird drift further and further out with the flock.

'This has to be the Sociable Plover … It can't be anything else,' I found myself saying out loud. 'This just has to be the plover.'

The light was fading. The flock was moving as if it was floating on the outgoing tide. The birds were getting further and further away. Panic was setting in.

'Please fly!' I urged the flock.

The flock didn't hear me. I stood in darkness and watched the mass of birds slowly fade into a distant indistinct blob on the mudflat.

The Sociable Plover was never seen again.

To this day, I am almost certain that the pale bird amongst the Lapwings that evening had to have been the Sociable Plover. It really couldn't have been anything else. But I can never be sure.

Of the birds I missed during 1985, I have subsequently managed to see them all, either in Ireland or abroad. All, that is, except for Sociable Plover. They are a globally endangered species and, despite searching for them at various locations within their global range, I have never seen one. Sociable Plover is one of the birds that I most want to see before I die.

Perhaps only when I encounter a Sociable Plover will my recovery from the trauma of 1985 be finally complete. For the moment, having written it down, all I can say is that I'm almost there.

Chapter 18 🔹

LIFE'S TOO SHORT

My notebook from the winter of 1986 contains some reasonable drawings and paintings, done in gouache and watercolours. I was very keen on drawing and painting back then. Flicking through the pages, I come across a nice painting of a Black-necked Grebe. It was based on several sketches I had done on 30 November 1986. I note with delight that I was travelling with Ronan that day. We were birding on the South Slob in Co. Wexford.

Since the dreaded autumn of 1985, I had been driving my own car, so Ronan and I weren't twitching or birding together as often as we used to. Ronan was also concentrating more on his photography than his birding. However, we were still in touch with each other by phone on a weekly basis. We were also serving on the Dublin Branch of IWC together. Ronan was one of my closest friends and a superb teacher when it came to photography; he had a great eye for a good shot. As I mentioned, he was a frequent visitor to my home where he always received a royal welcome from my family. His gentle nature, good sense of humour and sharp intellect made for superb company. Da and Ronan did not share the same political views and this made for many long debates between them. I would often leave them to it. Ronan was not much of a drinker, but he and Da would often sip a small glass of whiskey together. Ma would often include him in the pot for dinner if she knew he was arriving. He was always late, so more often than not his dinner was saved in the oven for him. Ma liked to feed him some good home cooking.

On this cold sunny day in November 1986, Ronan and I saw some great birds. As well as the Black-necked Grebe, we saw a Short-eared Owl and up to eleven Lapland Buntings. I was driving and I dropped him home to his house in Sutton that evening. Ronan loved his house, which he had bought a few years previously, as it was right beside the North Bull Island. In two minutes he could be on the mudflats watching waders and geese. He had even seen a Black Redstart in his garden. However, having a house was a burden on his budget. Few of us had even thought about buying a house back then; while we were off spending money on birding weekends and gear, Ronan was always saving to buy something new for his home. He had new furniture to pay for, a TV to buy out on a hire-purchase deal and a big mortgage to meet each month. For years we had spoken about doing a photographic wildlife safari in somewhere like Kenya, but each year he had this bill to pay or that loan to repay. While I envied him his lovely house, I didn't envy the financial strain it put on him. Life was too short to be worrying about mortgages and houses as far as I was concerned.

The birding year of 1987 got off to a good start, the highlight being a returning Whistling Swan to the South Slob in Co. Wexford. This North American race of Bewick's Swan was a real rarity (and still is) and had been present the previous winter at the same location. Ronan and I had twitched the bird together when it was first seen in Co. Wexford in the winter of 1985–86.

It was a source of good conversation when I next spoke to Ronan. However, this telephone conversation took an unexpected route. Ronan was keen to meet up to discuss something. He suggested that we meet for a pint, so a few days later we met up in Conway's in Parnell Street. It was 17 March, St Patrick's Day. The place was buzzing, but, as always with Conway's, there were nooks and crannies to find a quiet spot.

We sat catching up on the latest bird news and things in general. Then Ronan announced his big news. He had paid back a loan on some of his furniture and had even managed to finish off a hire-purchase on other domestic goods. He was now a little better off financially than he had been for a year or two and he had made a decision: 1987 would be the year of that Kenyan safari.

'What'd you think, Eric?' he asked. 'Are you on for it?'

I didn't need to be asked twice. 'You bet I am!'

'Great … I need to get more stuff for the house but I've decided to wait until next year,' Ronan said, adding, 'Life really is too short to put Kenya off for another year.'

I couldn't have agreed more.

The rest of the night was spent enthusiastically discussing our plans. We decided that we would definitely go to Kenya rather than to Tanzania and that we would visit some of the best reserves where we might manage to photograph big game species. We spoke about the shots we'd get. Then, of course, there were the hundreds of birds we'd see. Going on an African safari had been one of my burning ambitions since childhood and I couldn't wait. We decided that we would each select five wildlife safari companies and write to them for their brochures (we tend to forget how much easier it is these days to arrange such trips via the internet). We would meet again in two weeks. Hopefully by then we would have received a couple of brochures and these would allow us to develop our holiday plans. We left the pub that night to catch our respective buses home. We shook hands as we parted.

'It's going to be some trip,' Ronan said.

'Bloody right it will!' I answered. 'I can't wait.'

It was early morning on Wednesday, 25 March when I got a call from Gracer. To get a call at work from Gracer in early spring meant just one thing – a good bird had been found.

'Hi, Eric … How are you?' Gracer began. 'I have some news.'

'I thought you might have. Why else would you be ringing me? What's been found?'

'No, it's not bird news … Are you sitting down?'

His voice sounded very serious and it stopped me in my tracks.

'Yes … I am sitting down. What is it?' I said.

'It's Ronan,' he said.

'What about Ronan?'

Gracer paused to take a breath. He was obviously very upset about something.

'He's been found dead.'

His words hit me like a sledgehammer. It couldn't be true. Ronan was only thirty-one years old.

'Are you sure?' I asked.

It was such a stupid question, but I honestly thought that there must have been a mistake.

'Yes. We've all just found out this minute.'

Gracer worked in the same office as Ronan in the civil service and the staff had just been gathered together to be informed of Ronan's death. He had died in his sleep.

I simply could not comprehend what I was hearing. I thanked Gracer for taking on the role of the bearer of such news, put the phone down and left my office. I needed to take in some fresh air. I was in shock. Ronan was just five years older than me. This did not make any sense at all.

When Ma heard the news, she broke down and sobbed. Da shook his head, said nothing and left the house to go for a long walk.

At his funeral later that week, I spoke with his father. It was a terrible time for all of his family. His father shook my hand and thanked me for my friendship with his son. I told him of our plans to travel to Kenya later that year.

'Ah,' he said. 'That explains the neat pile of letters to holiday companies that were sitting on his kitchen table. They were all ready to be posted.'

Ronan was buried in Sutton, overlooking his beloved North Bull Island. I stood at his grave and wept for the loss of such a good, gifted, gentle friend.

The week following Ronan's funeral, I walked into Mercantile Credit Company in Dame Street in Dublin and applied for a loan of £5,000. This was an enormous amount of money in 1987; it would have been a chunky deposit on a house. Since I had a good job and a regular income, my loan application was approved.

With five grand in my pocket I went to a holiday company and booked a three-week holiday to Kenya and the Seychelles. I also bought seventy-five rolls of good slide film.

I departed from London on 31 August 1987 armed with bird and mammal books, bins, scope, camera, and rolls of film. We landed in Mombasa and travelled to Tsavo East National Park to stay at the Voi Lodge. When I was shown to my hotel room, I pulled back the curtain to reveal a sight that will stay with me for the rest of my life. My window overlooked a vast savannah where hundreds of African Buffalo, Zebra and Giraffe were wandering. Baboons were lazily hanging around the water holes just below the balcony, while Blue-naped Mousebirds perched on the nearby trees. Overhead, Bateleur Eagles soared while thousands of Little Swifts wheeled through the air. The large windows and the spectacular view gave the impression of watching a David Attenborough documentary on a giant screen. It was breathtaking.

On our first game drive that afternoon, we encountered our first Lions – a pride of six who were guarding a dead buffalo they had taken down the night before. Around the kill there were Rüppell's, White-backed and Hooded Vultures. There were Impala, Waterbucks, Grant's Gazelles and Oryx. This was just our first afternoon.

The safari took me across Tsavo East to Tsavo West, where I saw my first African Elephants. Seeing African Elephants was a childhood dream come true. It was here that I also encountered

Cheetahs and Leopards, two of the most beautiful cats in the world. The list of animals and birds goes on and on. I shall not elaborate except to say that the experience was beyond my wildest dreams. It would have been beyond Ronan's wildest dreams too.

We continued on to Amboseli, where the abundance of wildlife made Tsavo seem lifeless by comparison. The dry savannah and rich, lush wetlands were a wildlife paradise. Elephants and Black Rhino grazed beneath a backdrop of the snowy summit of Mount Kilimanjaro, while prides of Lions dozed under the shade of trees. We even saw the rare and elusive African Wild Cat here, with its beautiful markings and striking rufous-coloured ears. This, of course, is almost ignoring the birds we saw. At the water holes, African Fish Eagles were everywhere, while at one location we saw five species of kingfisher side by side. Goliath Herons fed alongside Saddle-billed Storks. Long-crested Eagles and African Hawk Eagles sat in the trees. We saw Anchieta's Sunbirds, Little Bee-eaters, Lilac-breasted Rollers, Superb Starlings, Crowned Plovers, Red-billed Hornbills – the list reads like a who's who of African birds.

At night, the sounds of calling Zebra, crickets and nightjars competed with the deep 'umm-waa' rumbling calls of male Lions. The experience was an assault on all of the senses.

The last section of the safari saw us drive north from Nairobi, up into the Great Rift Valley. The scenery was fantastic and, as we were driving over the mountains, our eyes were treated to the distant sight of a shimmering pink lake. This was Lake Nakuru. It was only as we came closer to the lake that we realised that it was pink because of the 500,000 Lesser and Greater Flamingos that fed there as one enormous flock. The rich wetlands of Nakuru and nearby Lake Naivasha were full of shorebirds, herons, egrets and kingfishers. Male African Paradise Flycatchers, with their incredibly long tails, darted around the trees, while Scarlet-chested and Tacazze Sunbirds added a dash of additional colour to the proceedings.

I spent my last night in Africa in a hotel room in Nairobi. The city was in darkness due to a power failure. Thankfully, the hotel had its own generator. I sat reading over my daily notes and logs, reliving moments and hoping that the shots I had taken would live up to my expectations. This had been an experience of a lifetime.

From Kenya, we flew out across the Indian Ocean (enjoying stunning views of Kilimanjaro from the plane) and landed in the tropical paradise of Mahé, the largest of the islands in the Seychelles chain. Flying into the airport, I looked down on the turquoise water and glistening white sands against the backdrop of forest-covered mountains … Well, there are few sights like it in the world. It was hot and humid when we landed. A small bus ferried us to the Reef Hotel, a short distance from the airport.

I walked into the small apartment that would be my home for ten days and was astounded by the view. The doors of the apartment opened out onto the beach; and there, just twenty metres away, was the Indian Ocean in all its glory. It was also strange to note that, as we were in the Reef Hotel, the letters 'RH' were everywhere. It was hard to forget that Ronan Hurley was meant to be on this trip too.

However, I wasn't here for a beach holiday: I was here to see some of the rare endemic species of birds that the Seychelles were famous for. Many of these islands have been isolated for thousands of years, so many of the birds have evolved into unique species that are found only here. Yes, a swim in the ocean was a daily routine both morning and evening, but I still had lots of film to get through and some great birds to see.

Birding on the Seychelles was not as productive as in Kenya, but it was a matter of quality over quantity. The tidal mudflats were full of shorebirds, while the forests and hotel gardens hosted some of those precious endemics like Seychelles Blue Pigeon, Seychelles Grey White-eye, Seychelles Sunbird and Seychelles Bulbul. Overhead flew Seychelles Cave Swiftlets. On the telegraph poles along the roads sat elegant Seychelles Kestrels. All of these species are special and each one was a real treat to see. Offshore, flocks of Wedge-tailed and Audubon's Shearwaters passed over the blue, calm seas, while, overhead, White-tailed Tropicbirds soared with their long, white tails trailing behind them like ribbons in the wind.

The privilege of encountering such rare species is a special thing. However, there was one place that I really wanted to see, a small, tropical island called Bird Island. This island is approximately 80 km north of Mahé.

Bird Island, as its name suggests, is a bird nirvana. It is about 1.5 km long and 500 m wide, and is home to almost one million nesting terns. It is a flat island surrounded by white sand and turquoise sea, with palm trees along the beaches. Only fifty people are allowed onto the island at any one time and their movements are restricted to certain areas. In the centre of the island is a large grassy area surrounded by palm trees. This is where the terns nest, all one million of them.

As soon as I had arrived on Mahé, I had enquired about the possibility of getting to Bird Island and was delighted to find that there was a cabin free on the island for one night during my stay. I didn't think twice. I booked the flight and the cabin. Landing on the small grassy strip of the runway was an experience in itself. The plane was a small eighteen-seater and the views from Mahé to Bird Island were spectacular.

The first birds I encountered were Fairy Terns nesting on the bare branches of the palm trees. These pure white, almost translucent birds have large black eyes (which make them so endearing) and a striking blue base to their black beaks. They were just sitting around on the trees, watching me (and the world) pass by. However, the island is famous for the colony of nesting Sooty Terns that occupy its centre. Amongst them was the odd Bridled Tern as well as Lesser and Common Noddies. On the beaches stood Lesser Crested, Crested, Little and Black-naped Terns, while Great and Lesser Frigatebirds soared overhead. It was tern heaven. It was a bird photographer's heaven too.

That night, I lay in my bed looking up at five Fairy Terns that were roosting on the wooden beams directly over my head. It was surreal to drop off to sleep with terns looking down at me.

I returned to Mahé with some wonderful memories and with some wonderful images. Bird Island had been such a great experience. However, with time now running out fast, I decided on one more adventure – to try to see one of the world's most endangered bird species, the Seychelles Magpie Robin. In 1987, this species was on the brink of extinction, with an estimated population of just thirty-five individuals. These were to be found on Frégate, a small, remote and densely forested island.

I made an enquiry about the possibility of getting to Frégate. The man I spoke to shook his head and smiled.

He told me it could only be done by using an air taxi, a small, chartered plane. 'You also have to pay a visitor's fee to visit the island,' he said.

He whistled in the way a workman does when he shakes his head and says, 'That's a big job!'

'All very expensive,' he said, shaking his head again.

'I don't care how much it costs. I want to make a booking,' I replied with a broad smile.

Life was too short to be in the Seychelles and not to at least attempt to see one of the world's rarest birds. Even if I got to Frégate, there was no guarantee that I would even find a Seychelles Magpie Robin, but I had to try.

And so it was that, on almost the last day of this wonderful three-week holiday, I boarded a tiny eight-seat plane and flew to Frégate Island. I was in the co-pilot's seat and the view from the cockpit was another memorable experience. The steep descent to the narrow grass runway was a little stomach churning, but flying over the hilly forested slopes with the anticipation of possibly connecting with this gem of a bird was worth it all.

The hilly part of the island is very much as it was when it was discovered by white settlers hundreds of years before. An old colonial house is the main building on the island and is a throwback to the days when parts of the island were used for growing coffee. The slave quarters are also still visible.

Walking along the tracks we were assaulted by the movements of Giant Millipedes that were 20 cm long. It was hard not to crunch their empty shells as you walked. Then there were the Seychelles

Skinks, endemic lizards that feed on the millipedes. Giant spiders in giant webs hung from every tree. Finally, giant Seychelles Tortoises, similar to those found in the Galápagos, ambled slowly around the tracks and the forest edges. These were the last remaining wild population of tortoises in the Seychelles; the islanders on Frégate did not like the taste of their meat and so these ancient reptiles survived into the modern era. The other interesting aspect about the island is that it is one of the few places where the famous Coco-de-Mer palm trees grow. Their striking separate male and female plants leave nothing to the imagination, while the nut they produce is the largest seed in the world. Being on Frégate was like walking back in time.

It was the presence of the Giant Tortoises along with pest eradication (especially of rats) that held the key to this being the last place on earth to find Seychelles Magpie Robins. The birds often fed alongside the tortoises, catching insects and millipedes disturbed by the grazing reptiles in much the same way our Robins follow a gardener in the hope of picking up worms that might be dug up.

I spoke with the guide on the island but he intimated that my chances of finding a Seychelles Magpie Robin were very slim at best. He told me that the birds were most often found in the more remote forests at the far end of the island. He gave me good directions to an area that would hopefully offer me some hope of coming across one.

Walking up the steep hills in the dark forest was draining. The heat and humidity were overbearing. The insects and millipedes were a constant distraction. I have to admit, I squirm a little in the face of things that have so many legs. Walking up a steep area of the forest, I came upon a small clearing where a stream ran across the track and down into the forest. Out of the corner of my eye, I caught a glimpse of a movement. I held my breath. Then, out in the open appeared this thrush-like bird with a subtle blue sheen to its dark plumage, and large, conspicuous white wing patches – a Seychelles Magpie Robin. I was watching one of just thirty-five individuals left on the planet at the time. I could hardly believe my luck.

It fed out in the open, tossing back leaves in search of insects in very much the same manner as a Song Thrush. I even managed to run off a few shots with the camera during the two minutes that the bird was on show. As quickly as it appeared, it disappeared back into the darkness of the forest.

I stood for a few moments to allow myself to take in what I had seen. It was a bird moment to remember for the rest of my life.

I returned to Mahé and spent my last night writing up my notebooks and logs. It was hard not to think about Ronan on that last night. As I remembered the many magical bird and wildlife encounters I had experienced, I knew that he would have loved every moment.

I came home to Dublin without a penny. However, I felt I was the wealthiest man on the planet. I had riches beyond my wildest dreams. I had memories. I had experiences. I had seventy-five rolls of film to enjoy and from which I could relive those moments.

It had been a three-week holiday of a lifetime, a lifetime that I now knew could be so short. Life really is worth living to its fullest. We all have a long sleep ahead of us.

Chapter 19

MISSING IN ACTION

There is one truth that all non-birders should know: when rare birds turn up, twitchers tend to forget about everything else.

In the past, I have forgotten to attend family dinners – but my family were well used to that. The problems always arose when it was someone else's family event. I soon learned to broadcast this truth. So, I always accepted an invitation to an event with the proviso: 'If a rare bird turns up, I probably won't be there.' It was the accepted norm for anyone who invited 'Eric the Twitcher' to anything. That was just the way it was.

This might seem very self-centred, but it was more that I had tunnel vision when it came to seeing rare birds. Perhaps I was single-minded, uncompromising; perhaps bloody obsessed is closer to the truth.

There was a strange logic to my thinking. Birding and twitching were my passion and seeing a rare bird that I might not ever see again was more important than attending the marriage of people I could see any day. I expected other people to understand this logic. In most cases, they accepted me for who I was, or perhaps my friends and family felt they had little choice but to accept me for who I was. One well-known British twitcher is renowned for leaving his own wedding to twitch a rare bird. I don't think I'd have ever gone that far, but I was never tested on that score.

In the height of the autumn migration season, birders and twitchers go into a world of their own. I would try to avoid any commitments during peak autumn months. I announced to everyone that, during September and October, they should either leave me off invitation lists or simply accept that I might be a no-show.

At home I explained to Ma and Da from very early on that if they had any consideration for my Irish list, then they should hold off dying during the peak autumn migration periods. It became a joke in our home. 'Don't die in autumn' was the golden rule.

'If your father dies, we'll stick him in the freezer until November,' Ma would quip. 'But we're going ahead with his funeral without you if he starts to go off!'

Da gave me a guarantee that he'd try to keep alive during the autumn months so as to avoid causing any conflict of interest.

'I'd hate to be the cause of you missing a good bird,' he'd say. Then, with a sad face and a sad tone, he might add, 'Ah, never mind your poor aul' Da. I'll lie dead in the freezer until you've seen your rare birds. Don't worry about me … I'll be okay in the freezer.'

However, while my family accepted my frequent absences when it came to important events, not every birder's family was quite as experienced as mine. One particular case comes to mind from August 1988, when Michael O'Clery and I went missing in action.

I picked Mick up at his home very early on Saturday, 6 August. He was exhausted. He had been very late coming home the previous night and now we were heading off in the early hours of the morning. We headed for Co. Wexford and spent the day birding around the hotspots. It was a warm sunny day but the best we saw was a single Yellow Wagtail. We didn't mind as we already had plans set for that evening. We were killing time before heading to Co. Waterford. That night we were joining a group of 'invited' birders to see our first Irish Nightjars. Now, this was the kind of invitation that we could always accept.

We gathered in a pub near Dungarvan and met the gang. At the designated time, we went in convoy along back roads, laneways, tracks, highways and byways until we eventually swung into a remote woodland carpark. As we all swung into the carpark, we frightened the life out of a 'courting' couple who were enjoying the warm summer evening (they must have been warm as it seemed they had very few clothes on). The poor couple. One minute they were alone in this remote little woodland enjoying some intimate moments, and the next they were surrounded by five cars with headlights on and then by about twelve birders with bins, scopes and cameras.

It was a lovely walk into the woodland. There was such a spectacular sunset that even the most cynical of birders in the group had to admit that it was beautiful. We walked for fifteen minutes, got into our positions along an open track and waited. Then we heard that most wonderful sound – a churring Nightjar. It was the first time I had ever heard this sound in Ireland. In fact, I have only heard singing Nightjars in Ireland on a handful of occasions since that night.

We waited (not very quietly, I hasten to add, as everyone was still laughing about the couple back at the carpark) until the churring stopped. This was followed by the 'thiock' call of a flying Nightjar. We held our collective breath and then right over us flew this long-winged, graceful bird. It had big white patches on its wings. It even landed on a tree right beside us. It was a moment of magic and was a vital part of my rehabilitation from missing that Hook Head Nightjar three years earlier.

We saw two Nightjars that night. It was a superb experience. As we gathered back at the carpark, some of the birders announced that they were heading on to Ballycotton (BallyC) in Co. Cork, where they would sleep in their cars and resume birding in the morning. Mick and I headed back to Dublin. Mick had a family gathering that he was due to attend on Sunday so we had to get back. We reached Mick's place at 3.30 a.m. It was after 4 a.m. when I got home. I fell into bed exhausted.

I was awoken in the late morning by the phone ringing. It was close to midday. I answered the phone yawning, but I was very quickly brought to full alertness by what I heard.

'Eric, get your arse down to BallyC. There's two bleedin' Caspian Terns, a Stilt Sand, and Kelly's after finding a fuckin' Least Sand!'

I stood in shock, trying to take in the news. I had seen my first Irish Stilt Sand only a few weeks ago. Now there was another, but it was only the supporting act to two mega Irish ticks. Least Sandpiper is one of the rarest Yankee waders in Ireland. To see that along with two Caspian Terns would be unreal. I was plunged into automatic twitch mode. Adrenalin pumped through my body and my mind processed what I needed to find: bins, scope, car keys, wallet.

I cursed the fact that we had decided to drive home the previous night. But I decided to ring Mick in the hope that I might be able to get him before he went to his family gathering.

I rang his number. It rang out. There was nobody home.

'Feck, it's too late,' I said to myself. 'He's already gone.'

I decided to ring his number again. This time a weary Michael answered the phone.

'Hello … eh … the O'Clerys. Michael speaking.'

'And very feckin' polite you are too … Now listen up. There's two Caspian Terns and a Least Sand at BallyC. I'll be at your place in about thirty minutes.'

Then I remembered that he was meant to be going somewhere.

'Is that family thing on today? Can you get out of it?'

Mick laughed. 'They obviously didn't know I was home,' he said. 'In fact, they haven't seen me since Thursday morning, so they've gone without me.'

That was great news. He was free to twitch. I picked him up and we headed south as fast as we could. Actually, we headed south as fast as we dared, because my brakes were not great. I had planned to get them fixed the following Monday.

We safely arrived in BallyC at 5.45 p.m. and met some of the birders. We were buzzing with excitement and quickly got the information we needed.

'The Least is on Shanagarry. The Stilt and Caspians are down at the lake.'

We stood for a moment. Which direction would we go in?

'I'd go for the Least first,' a birder suggested. 'The Caspians are giving themselves up and look settled down at the lake.'

So we headed into the back of Shanagarry Marsh to look for the Least. We found it very quickly. It was a thing of beauty (the word 'beauty' for wader fans has a totally different meaning from how many understand the word). It was like a mini-Pec Sand. It was a gem of a bird.

With one great bird under our belts, we ran off down the beach to the lake. There were several birders there as we approached, which is always a good sign. However, we were greeted with bad news.

'The Caspians have flown out into the bay. They're not at the lake any more.'

I couldn't believe it.

'Na … don't worry. They've done that before and came back in,' we were assured.

We found the Stilt Sand and enjoyed good views of it but all the while we were keeping our eyes peeled for these enormous, red-billed Caspian Terns. It wasn't as if we could miss them. We stayed until darkness set in, but they never showed up.

Meanwhile, back in the O'Clery household, Mick's mother, Della, was getting very worried about her eldest son. He had headed out on Thursday morning. She knew he was staying with friends that night so had expected to see him home on Friday. But he had never come home. She rang his friends and they told her that he had stayed on Thursday night and had indeed left them on Friday morning.

Whatever about him not coming home on Friday night, it was now Sunday night and there was still no trace of him. As all good mothers do, she began to fear the worst. Of course, what she didn't know was that he had been home twice since then but no one had seen him. Now he was in Cork. This was of course before the days of mobile phones. Mick was suffering the trauma of missing the Caspian Terns, and it didn't dawn on him to ring home from a phone box. Missing Caspian Terns was far more important than a missing Michael.

As darkness set in over BallyC, we faced the reality that we had made the wrong choice. We should have gone for the Caspians first and the Least Sand second. Hindsight is a marvellous thing. We resigned ourselves to the fact that we had dipped. As I started the car and headed along the back roads towards the main Dublin road, the brakes on the car seemed to be lacking in sharpness (isn't that a great way of saying that they seemed to be failing?). As I approached the main road, I went past the turn-off. Mick realised I had not taken the correct road.

'Eh, that's the turn off, Eric. You've just missed it!'

'I know,' I replied. 'I didn't miss it. I just couldn't slow down to take it … Me brakes are gone.'

It was too dangerous to drive any further. We would kill ourselves if we tried to drive all the way back to Dublin, so we drove very slowly towards the outskirts of Cork, where we stayed overnight. Tomorrow we would get the brakes seen to. If the Caspian Terns showed up again, at least we'd be in a good position to 'run' for them.

Monday morning, 8 August, arrived and Della was now really worried. She had hoped to see Mick home when she got up that morning, but there was still no sign of him. Where was he? Even worse, she knew that he was due to go to college that morning to confirm that he would be attending for the following term. She consoled herself with her belief that Michael was a responsible young man and that, if he was off gallivanting, he would at least remember to look after the college business himself. Of course, what Della didn't realise was that college was the last thing on Mick's mind. He had two Caspian Terns to think about. He had not only forgotten that his absence might be causing some concern for his mother, but he had also totally forgotten about college as well.

Meanwhile, I was on the phone ringing work. I told them I was dying of a stomach bug or a fever, or something along those lines. Basically, I was too sick to go into work and I didn't know if I'd make it in for the next few days (or, in other words, I didn't know how long it would take to see the Caspians). With work sorted, it was time to get the brakes fixed.

That Monday, we waited around all day while the car brakes were being fixed. Throughout the day, we phoned Cork birders to see if there was any update on the Caspians.

'No, Eric, I'm afraid there's still no sign of the terns,' was the answer on more than one occasion.

As my now safe car was returned to me late that evening, I decided on one last throw of the dice. I rang a BallyC birder one more time. Dennis answered.

'Eric, I'm so pleased you rang again. They've just turned up!'

It was the news we had prayed for. We arrived in BallyC in darkness. The night ahead would be a nervous, sleepless one for us.

Back in the O'Clerys, Della was now up the walls with worry. That evening, she had taken a call from the college. They were looking for Michael. He had not shown up for his appointment. While Michael was known to go off on occasion, this was surely very different. It was Monday night and the last sighting of her son had been on Friday morning when he'd left his friend's house. She was now convinced that something terrible must have happened to him. She was about to ring the Guards to report him missing but decided that it might be an idea to ring me first to see if I could shed some light on his whereabouts. Perhaps I might even have spoken to him in the past few days.

She phoned my home and Da answered.

'Hello, Mr Dempsey. It's Della here, Michael's mum. Is Eric there?'

'Eric? No … I'm sorry, he's not here. Can I take a message?'

'Eh, no thanks … It's just that I'm worried sick about Michael,' she said. 'I haven't seen him in days and no one knows where he is. When Eric gets in, could you ask him to ring me please?'

'Oh, I wouldn't worry too much,' Da replied. 'Sure I haven't seen Eric in days either. I think he's off birdwatching somewhere.'

'And is Michael with him, do you know?'

'Yes, as far as I know he might be. I think there might be a few rare birds in Cork. He came home very late on Sunday morning, and sure he was hardly here before he was off again.'

Now the pieces of the jigsaw were beginning to fall into place in Della's mind.

'So, I wouldn't worry if I were you,' Da continued. 'I'd say the pair of them are down there.'

Da's years of experience told him that if I went missing during the migration period, there really was nothing to worry about. His words put Della's mind at ease. Her beloved birding son wasn't lying somewhere in a ditch; he was twitching.

Tuesday, 9 August dawned bright and warm. We slept in the car and were out around the lake as the sun rose. There before us was a pair of exquisite Caspian Terns. They stood among the gull flocks and flew with long, powerful wings around the lake. We watched them for hours. We then returned back to Shanagarry, where the Least Sand was still on show. It was a very special BallyC morning.

Late in the afternoon, we headed back to Dublin. It was early evening when I dropped Michael off home. We were met by a very irritated but very relieved Della.

'Where were you, Michael?' she asked. 'Had you forgotten that you were meant to be in college yesterday? Could you not have phoned me? I was worried sick about you!'

Before Mick could answer, I intervened.

'But, Della, we got three ticks,' I said with real conviction.

I smiled at her.

'They were seriously mega birds,' I added.

There was no answer to that. This was the truth.

She looked at me and then looked at Michael, and she laughed.

'Och, you boys!' she said, shaking her head and rolling her eyes to heaven as she walked away.

You might say that this was the moment when she accepted the undeniable truth: during the autumn migration months, birders *will* go missing in action.

Chapter 20

ON SOUND ADVICE AND MEGA BIRDS

I have to admit, I'm not a great fan of camping. I remember Da helping me to pitch a tent in the back garden when I was a kid. I planned to stay outdoors for the night. I zipped up my tent and settled in. Within thirty minutes, it began to rain heavily, and within an hour of turning off my torch for the night, I had abandoned my sodden tent and was back in the comfort of my own bed. Yes, when I was in the Cub Scouts, the idea of going off to the Dublin Mountains to camp held a greater promise of adventure. However, I have long since lost that sense of adventure about camping. Let's just say it has taken a beating from the Irish weather.

However, my dislike of camping was not in my mind as I set off birding for a long weekend on 31 August 1989. I had taken Friday and the following Monday off work. I picked Mick up at his place on Thursday evening and we set off down to Co. Wexford. It was a calm and clear night as we pitched our tents between the East End and the Forgotten Corner of Tacumshin Lake. We had eaten 'grease' on the way so there was no need for making dinner. For the uninitiated, 'grease' is the term used by Irish birders to describe food from chippers, or greasers. We spent the last few hours of daylight birding around the lake. We saw nothing.

I slept reasonably well, despite the cold, and, having had a quick breakfast, we headed off with high hopes. We walked and birded for hours. Apart from one Little Stint and a Marsh Harrier which had been present for weeks, we saw nothing. We were a little demoralised. We had the whole of Tacumshin to ourselves on the first day of September and there was nothing to be seen. So, in true nomadic birding style, we packed up and moved on.

We decided to move west and opted for Ballycotton in Co. Cork. We arrived late in the evening, pitched our tents in the dunes near the beach and went into the village. We had a great night in the Anchor Drop pub, run by the O'Sullivans, BallyC's birders-in-residence. The walls of the Anchor Drop were adorned with images of the Pied Wheatear from Knockadoon, an Ivory Gull that had been seen at Ballycotton Pier in January 1980 and Ireland's first American Coot, found at BallyC in February 1981. The pub had a great atmosphere and was the centre of bird information when away from home.

We left the pub and settled down for another night under canvas. It was a warm night and I slept well, perhaps helped by the few pints I had consumed.

With renewed enthusiasm, we set out at first light to check the pools, lake and beach, and then the beach, lake and pools. We checked every square inch of the place ten times over until at last we found a good bird. Well, it sort of found us. We were returning back from the lake for the fifth time when a bird flew over us. Only for the fact that it called, we might not have even noticed it. The call was a crisp, clear 'kyep'. We looked up to see a Long-billed Dowitcher flying over us. It was obviously fresh in. It circled around the lake and went into the back channels.

It gave us hope, so off we went checking the whole place again. We saw nothing else and, worse still, we didn't see the dowitcher again either. We abandoned our birding in the early evening and headed for Cloyne, the nearest village, for grease, only to find that the greaser didn't open until 9 p.m. Even the chippers weren't on our side. So, we ate in a good restaurant by the pier in BallyC that evening, and once again we enjoyed a great session in the Anchor Drop.

The O'Sullivans kept a logbook and encouraged all birders who visited their pub to write something into it. It was filled with sketches and comments. Writing into this book became one of the great traditions of birding. Over several years, the logbook became a treasure trove of comments, descriptions and sketches made by a whole range of visiting birders. I really would love to see that book now. It would be like reading over a little bit of birding history.

We fell out of the pub in the small hours of the morning and made our way back to our tents. I slept like a log (there is a pattern building here).

Sunday morning was warm and bright. However, the forecast warned that strong westerly winds and rain were due to hit the west coast by late afternoon and evening. We set off birding around BallyC, checking the lake, the beach and the pools, and then the pools, the beach and the lake. BallyC was dead.

We had drawn a blank. It had been a hard three days' birding and camping, and we had seen very little. Yes, we had found a Yankee wader, but even that dowitcher hadn't exactly co-operated with us. With bad weather forecast, the thought of camping for another night did not exactly appeal to me. Then we made a decision that would make this trip one to remember.

'Let's head to Kerry!'

So, off we went, birding nomads in search of rarities. It was a long haul over to Co. Kerry but the conviction that there must be some good birds to be found somewhere kept us going. We birded at Blennerville but there was nothing there. We birded at Black Rock. There was nothing there. As we reached the village of Ballyheige, the rain began falling and the wind picked up. We drove along the small sandy track between Akeragh Lough and the sand dunes, found ourselves a flat piece of ground and pitched our tents. We went back into the village, bought some grease and beer, and returned to our 'encampment'.

The wind speeds picked up during the night and the rain fell. Thankfully, where we had chosen to camp offered us some shelter from the wind. I awoke early. The rain had stopped and the sun seemed to be shining.

It was then I heard a noise just outside my tent. I saw a shape going by, rubbing up along the side of the tent. Then another.

'Ah, for feck sake … What are ye doing?' came a voice.

'Baaaaaaa,' came the answer.

'Holy Mother of Jesus … would ye ever feckin' get outta there … feck yis!'

'Baaaaaa … baaaaaa.'

I heard the zip of Mick's tent being undone. Obviously Mick wanted to see what was going on.

'Hello,' I heard him say.

'Oh … hello … I didn't see ye there!'

I unzipped my tent to see a man surrounded by about thirty sheep.

'Ah, Jesus Christ…would ye look at that one,' he said, pointing out one sheep who had managed to find a small gap in a fence and was off up over the sand dunes with another following quickly behind.

'You're having trouble with the sheep?' Mick suggested.

Between the shouting and cursing, he told us that the sheep had broken out of a field and he was trying to bring them down the track to where they belonged. They had other ideas.

He looked at us.

'Let me give ye some sound advice,' he said.

I sensed that he was about to impart some hard-earned wisdom from years of worldly experience.

'If ye want to get to know women, get yourself some sheep. They're so feckin' contrary. They do what they want, when they want!'

He looked on in despair as all his sheep followed the first one over the dunes and out of sight.

He sighed.

'I'm telling ye, if ye can understand sheep, ye can understand women!'

With that he was off after his sheep, cursing and shouting as he went.

It was an unusual start to the morning.

We decided to have a quick look at Akeragh before breakfast. It would be many hours before we'd eat …

As I contemplated the advice bestowed upon us earlier, I almost walked on a small wader that was scuttling about in the long wet grass beside a pool. I stopped. It disappeared behind some reeds and emerged with another bird. They were both Pectoral Sandpipers. I was delighted. At last our luck had changed. We enjoyed super views of these little gems before moving on.

We separated and started checking sections in the middle of the lake. There was a small area of reed-fringed mud and I saw a lovely Water Rail feeding out in the open. It is rare that they allow such good views, so I took time to enjoy this colourful bird. Then, just behind the Water Rail, I caught a glimpse of another rail-type bird. It was about the same size as a Water Rail … but was that a short yellow bill I had seen? It had gone back into deep cover so I couldn't be sure.

I called Mick over.

'I'm sure I've just seen a Spotted Crake,' I whispered to him.

We stood for over thirty minutes. All we saw was an adult and an immature Water Rail. I was sure that I had not mistaken the young rail for a Spotted Crake.

I moved on while Mick decided to stay watching the area a little while longer.

I started checking through a small flock of Dunlin in the hope of finding something amongst them. There was nothing of note except a single Ruff. With thoughts of breakfast now taking over, I decided to head back. I was walking away from the lake and could see our tents near the dunes in the distance. It was then that I spotted something high in the sky, way out to sea.

I raised my bins. It was a bird and it was flying straight towards me. I couldn't make out a thing; it was just a faint moving blob. It was flying fast. I kept watching it. Slowly it began to take shape. It looked like a wader. I kept watching as it came straight for Akeragh Lough. It was descending but still flying very fast.

I stood still as it came closer and closer. It flew right over the lake and veered left up a channel out of view. I could hear the wind rushing through the bird's wings as it passed overhead. As it flew up the channel, I thought I had seen a white rump and long legs. Perhaps it was a Curlew Sandpiper? I retraced my steps back to the lake and very slowly walked out to get a view up the channel. At the top end I could see a medium-sized wader. It was standing in the water and was very agitated. It was bobbing and looking around. It was nervous. I was afraid my presence might frighten it. I remained still, not even lifting my bins.

After a few moments, it seemed to settle down. I raised my bins to feast my eyes upon the best bird I have ever found. There, just in from a trans-Atlantic flight, was a summer-plumaged Stilt Sandpiper. I couldn't believe my eyes. I took a look through the scope to convince myself that I wasn't hallucinating. I wasn't. It was a long-legged, chestnut-faced, barred North American beauty.

I looked to see where Mick was. He was still searching for the Spotted Crake. I waved frantically to him and he was beside me in a minute. We both watched this bird for over an hour. It had started feeding and was now showing very well. This was a major bird. It was only the seventh time the species had been seen in Ireland. Not only had I found it, but I had also the unique privilege of actually watching it arrive in off the sea.

I left Mick at Akeragh and dashed back to the car. I drove into Ballyheige, found a phone box and put the news out on the grapevine. Dempsey and O'Clery were having a good day: two Pecs and a Stilt Sand at Akeragh. I also bought some food and a few cans of Coke before driving back. When I arrived back at our tents, I could see Mick was still out in the middle of the lake. He saw me arriving. As soon as I stepped out of the car, he began waving.

I threw the grub into a bag and walked as fast as I could to where he was. I was panting when I reached him.

'What have ya got?'

'A Spotted Crake!' he replied.

We waited and watched. A few minutes later, out walked a Spotted Crake. This was a tick for me. It was everything I hoped a Spotted Crake would be. It had the short yellow beak with a red base, and dense white spotting on its grey underparts. The brown back had big black spots. It walked with short, deliberate steps on thick green legs and long green feet. It was truly a bird to behold.

We munched on our sandwiches and drank our cold cans of Coke. A cold can of Coke never tastes as good as when it is drunk in the presence of a Spotted Crake and a Stilt Sandpiper. It was a day to savour. Our efforts and failures of the previous days were all worth it for this moment.

After hours spent enjoying our birds, we took down our tents in the late afternoon and left Akeragh for Dublin. Our expedition was over. Our summit attempt had been successful. I returned to base camp in Dublin a wiser man.

A wiser man?

I learned that it pays to never give up, to keep going even if you think there isn't a hope of finding something.

I learned that camping in Ireland isn't so bad after all.

Most of all, I learned that some people just aren't cut out to be sheep farmers.

Chapter 21 ➤

THE RISE AND FALL OF THE BIRDLINE

The 1980s ended very well for me. In November 1989, I saw my first Irish Blue-winged Teal (from North America) at Tacumshin Lake in Co. Wexford. It was a grotty female type and it certainly was not the most attractive duck I had seen. However, this was my 300th Irish tick. I had crossed the Rubicon of Irish birding and joined the elite in the '300 Club'. Since I had ticked off my 200th Irish bird, the Pied Wheatear, at Knockadoon Head in November 1980, it had taken nine years of hard twitching to see the further 100 species.

As the decade ended and the 1990s arrived, it seemed that there were more and more birders out and about. It was great to know that there were so many people keen on seeing birds. Of course, that also meant there were lots of newer birders out finding rare birds. The real problem was that these new birders were not always in the grapevine. For us, 'the chosen people within the inner sanctum' of the grapevine, this spelt trouble ahead because we had no way of discovering what these new birders were finding. If someone found a mega in Co. Cork and was not in the grapevine, then that news might not filter through to us for days or even weeks.

The grapevine worked very well for many years with the number of active birders and twitchers at around thirty to forty. I was lucky to have been within the inner sanctum for over ten years. But all it would take would be for one link in the chain to be broken and the whole grapevine system would fall down. This began to happen in a small way during the last months of 1989, and by the spring of 1990 it was happening a little more frequently. Despite these small glitches, the grapevine was still the only source of bird news and, so far, we had not missed any major rare bird.

All that changed when I came to work on Monday morning, 2 July 1990. I had been away at a relative's Confirmation. As I have already mentioned, I had always tried to resist attending anything of this nature – Confirmations, weddings, even funerals – in order to avoid, for instance, having to inform the bride and groom that a rare bird had turned up and that I was off twitching. It was a well-known fact that in autumn, the only funeral I might attend would be my own. However, July seemed a reasonably safe month to risk attending a non-birding event. How wrong was I?

I sat down at my desk and was reading through work files when the phone rang. It was Gracer. He was in good form.

'So, did you get the tern?' he asked.

'Tern … What tern?'

I didn't like the sound of this.

'The Gull-billed,' he replied.

'What Gull-billed?' I asked with real urgency.

'The Tackers one …'

'There's been a Gull-billed at Tackers?' I asked with real incredulity.

'Eh … yes. It's been there since Saturday. Did you not hear about it?'

I paused before I answered.

'There's been a Gull-billed at Tackers since Saturday? Jesus Christ … this is the first I've heard of it!'

'Feck!' Gracer said. 'I thought you'd heard about it. We were wondering where you were. It was showing well on Saturday and all day Sunday.'

I couldn't believe what I was hearing.

'I was sure someone said that they had let you know …'

Gracer was beating himself up but it wasn't his fault. I was away from home so was away from a phone.

'I'm sure it's still around,' he said. 'Why don't you ring Dave to see if he knows.'

I thanked Gracer and hung up. I then phoned Dave in Wexford. His wife answered the phone.

'Hi, it's Eric here. Is Dave there?'

'No, Eric. You're probably looking for information about the Roller, are you? I don't think it's been seen in the last hour.'

'What Roller?'

'The one Mattie found near his house this morning.'

I thanked her for the information and hung up.

Holy Jaysus, a Gull-billed Tern and a Roller in Wexford! What was going on?

I rang Gracer back.

'Just spoke with Dave's wife,' I told him. 'He's not at home. He's gone to look for the Roller.'

'Roller? What Roller?' said Gracer.

'The feckin' Roller that was seen in Wexford this morning!'

After a number of frantic phone calls, I was on the road that afternoon. Alas, there was no sign of either bird. A superb adult Pacific Golden Plover offered a good consolation prize but did not make up for missing two Irish ticks. After speaking to other birders, it seemed that everyone thought that someone else had been in touch with me about the Gull-billed Tern. No one in Dublin knew about the Roller until I got the news out. The grapevine was collapsing.

As I was leaving Wexford that evening, the discussion centred on how to resolve the grapevine issue. There were too many birders now in Ireland and the grapevine simply could not function properly. The ideal solution would be to have one central source for getting news and for reporting news, bypassing the grapevine system totally. In Britain, some birders had started the Birdline, which gave everyone access to bird news for a small fee. If only someone would do something like that in Ireland …

By the time we arrived home from Wexford that night, the concept of an Irish birdline was born. Over the following days, after many hours of discussion, the Birds of Ireland News Service (BINS) was born. By the middle of July, I had BINS registered as a limited company, with Mick listed as one of the directors. By then, I had had discussions with a technology company that handled what were known as 'premium-rate phone services'. Such services were better known in 1990 for sex lines, but thankfully not this company. Instead, it offered the likes of weather forecasts and stock market information. It was a perfect fit. By the end of the month, I had done everything that was required for the introduction of the BINS line. We were all set to kick off the service by the beginning of August.

The concept for BINS was easy. It would provide one central source for all bird news. If anyone found a good bird, all they had to do was to make that one phone call to BINS. The BINS line would be updated with that news, to which everyone would have access. There would be a small fee to get the news, of which BINS would receive 25 per cent.

The charge for getting the bird news from the BINS line was set at less than 60p (90c) per minute, but it was broken into one-second slots. So if a birder rang the line and stayed on for twenty seconds to get the latest update, then they only paid 20p. We agreed this before BINS was launched as we felt that birders might only stay on for very short amounts of time just to hear the latest information, and so they should only be charged for the actual time they spent listening to the news. Unlike everything else in Ireland, the cost of accessing the BINS line never increased in twenty years.

The BINS line would be updated throughout the day, with a full update of all the day's news taking place at 9.30 p.m. every night. This meant that birders who were not in the inner sanctum of the grapevine would now have access to all the news.

For some who enjoyed the elitism of the inner sanctum, this was perhaps an unwelcome development. The very idea of all birders (including beginners) being so easily let into the loop of rare-bird sightings did not appeal to them.

All of this was the theory, but in practice we needed to set several things up. The most important thing was to have a phone number that people could ring with their news. None of us had our own home. By early 1990, I had moved out of home and was sharing a house with a gang of people. I was now living in Glasnevin, which was no more than five minutes' drive from good home cooking once or twice a week. It was not an option to approach my landlord with the idea of using his phone number for the BINS line. This was a major stumbling block.

In mid-July, when I was up at home, I told Ma and Da about setting up the BINS line. They were very interested.

'What's needed for this to happen?' Ma asked.

'Well, there are two basic requirements,' I explained. 'The first is to get an answering machine in order to take all the calls. But once I have that, I then need to get a telephone number to attach it to!'

Ma and Da looked at each other and they seemed to read each other's minds.

'Would our home phone work for you?' Da asked.

I explained to them that having the BINS answering machine attached to their home number would be a great inconvenience to them. Birders would be ringing at all hours of the day and night with news. If their friends rang them, they would be met by the BINS answering machine first.

In the true spirit of Ann and Tom Dempsey, they supported our little venture and allowed their own home phone number to become the main BINS line to which all news would be reported. Not only that, but they went out and bought me a top-class answering machine, one that could be remotely accessed.

'Let's just say it's our contribution to your new venture in life,' said Da on their behalf.

I could not believe their generosity. In fact, that is not true: this was typical of their generosity. I could not believe how lucky I was to have parents such as these.

That was one major problem solved. Now we had to entice people to actually phone in with their news. With that in mind, we decided that, while we're setting up the BINS line, we might as well publish a new quarterly journal. People would subscribe to this journal for £10 per annum but would receive a discount for every piece of news that they reported to the BINS line. For some, this would mean that they could receive the journal the following year without having to re-subscribe.

So, we began the task of creating a brand-new bird journal which would include everything to do with Irish birding. It was the first Irish birding journal for Irish birders (and those interested in Irish birding). Mick would be the designer. I was one of the editors. We even brought on board a group of top identification consultants, including Jim Dowdall, Anthony McGeehan and Killian Mullarney. With such a strong team behind us, we were ready to launch *Irish Birding News*.

With everything now in place, we launched the BINS line onto the Irish birding scene. We had a Stilt Sandpiper as our logo (it was a special bird for us) and we created a slogan: 'Would you go birding without BINS?'

On 10 August 1990, the first official BINS evening update took place. It featured news of a Subalpine Warbler on Cape Clear and a White-rumped Sandpiper in Ballycotton. The Irish birdline was born.

In September 1990, the first issue of *Irish Birding News* was published and posted out to our small number of subscribers. It was received well by the Irish birding family. At last, we had an Irish journal dealing with Irish birds and birding. The journal was published for four years (four issues per year), but in the end we had to wrap it up. There were not enough birders in Ireland to subscribe to such a journal. Even though the contents were well received here in Ireland, and amongst our European and American subscribers, we struggled to cover the costs of each issue. We soon realised that the cost of publishing the journal versus the revenue earned from subscriptions and the BINS line simply meant it was not viable.

It was also disheartening to learn that at least one group of ten Irish birders decided to put £1 each into a kitty and subscribe for just one person. They then photocopied copies of *Irish Birding News* for the group. Despite our best efforts, we could not keep our journal going. When the last issue was published, we were sad that Ireland had lost its first and only bird journal. However, we were also immensely proud of our efforts.

The BINS line, on the other hand, took on a life of its own. That first autumn, I was updating the line many times a day. News poured in each day, and Ma and Da would hear it as it broke. Each person who had subscribed to the journal was given a membership number. One elderly but very active birder was the only person to ever quote his number. He became known as '007' at home. Da might say something like '007 had a good day yesterday' or 'How is 007? We haven't heard from him this week.'

That October, an Ovenbird was found on Dursey Island (that remote piece of land in the Atlantic Ocean off south-west Co. Cork) by Tony Lancaster and Kieran Grace. This was a mega bird and I scribbled the details down. I picked up the phone to update the BINS line and for some reason said the following:

'This is a BINS update at 3.30 p.m. on Monday, 24 September. We have a RED ALERT in operation … repeat … a RED ALERT … an Ovenbird has just been found …'

It was the first of many 'red alerts' over the years and I have fond memories of many rare birds being reported to me. There was even a 'double red alert' with the finding of a Solitary Sandpiper and a Northern Waterthrush within moments of each other on Cape Clear in 2008. It was the only double red alert ever announced.

However, what I remember most are the moments when I was contacted by excited and shocked birders who had just been skilful and lucky enough to find a mega. These were always great moments, sharing truly magical phone calls with these birders. Of all of these, there is one moment that stands out above them all.

I received a call one afternoon from my good friend Maurice Hanafin. It was Sunday afternoon, 8 October 2006. It was a conversation I will never forget.

'Eric, it's a mega!' he said.

'Okay, Maurice. What have you found?'

'It's a mega, Eric … a mega!'

'Okay, Maurice. Where are you and what have you found?'

'I'm at Kilbaha. It's a mega!'

At this stage his phone was taken from him, and the calm voice of his friend Seamus Enright took over.

'Hi, Eric, Seamus here … Maurice is a bit excited. He's just found Ireland's first Canada Warbler,' he said.

I wouldn't blame Maurice for being in shock. This was not only a mega bird but a gorgeous bird too. I was delighted for him. Hundreds of birders travelled to Kilbaha, near Loop Head in Co. Clare, to see his Canada Warbler.

It is worth remembering that, for the first six years, all of these BINS updates were done without the technological luxury of mobile phones. If I was near a touch-tone phone at home (or at work), I would ring my parents' number, key in a code to listen back to the news on the answering machine and then update the line with the latest news as it broke. The whole day's news would then be given in order of rarity in my 9.30 p.m. update.

However, if I was out birding when news broke, I would then need to find a telephone box and queue up; if the phone was not a touch-tone one, I would have to use a small gadget that mimicked the tones (sounds) of a touch-tone phone, listen to the news on my answering machine and then update the BINS line. My heart was broken on so many occasions as I tried to find a phone box that worked. It seemed that every second phone box in Ireland was out of order; in some places every phone box was broken.

There is one very special phone box moment that I remember well. It was 12 September 1993. I was in Co. Wexford and there were storm-force winds and heavy rain all day. I found my way into Lady's Island Lake in search of a working phone box. There was one just near the toilets in the village. As I stood inside trying to make my calls, the box was almost lifted off its foundations and the door blown off its hinges by the force of the wind. I checked my answering machine and pressed my little ringtone gadget to listen to the news. I was astounded to learn of a Temminck's Stint – at Lady's Island Lake.

I updated the BINS line, left the phone box, walked fifty metres and got a tick. It had to be the shortest twitch on record.

In 1994, I eventually bought my own house and the BINS phone number was moved to my new number. Poor old Ma and Da had spent four years listening to bird news every hour of every day. They were so supportive, but I am sure they were more than happy 'to see it off the premises'. For a good while afterwards, people who still had their number would ring their house with bird news. They were both dab hands at taking detailed bird news for me.

Then, at last, came a big moment. It was 14 September 1996. I was standing in the Wheatear Field on Cape Clear Island and I did the first BINS update using a mobile phone ever. The news that broke that afternoon was of a Greenish Warbler in Muller's back garden in Wexford. Having a mobile phone made running the BINS line so much easier. I could now update wherever and whenever I needed to.

From the start of the BINS line in 1990, I had become the epicentre of bird news in Ireland. Birders from all over Ireland were contacting me with information. I was also the main contact for British birders keen to twitch Irish birds. With this came a pressure to be up to date with everything and to know what I was talking about. I recall on one occasion taking a call from a well-known British twitcher. I was in Co. Kerry watching Ireland's first Brown Shrike. It had just been confirmed as a Brown Shrike after a discussion between the many birders present. My phone rang and I answered it.

'Hello, Eric. Eh, just wondering if you can tell me if that shrike is confirmed as a Brown Shrike?

'Yes, it is a Brown Shrike,' I replied.

'Have you seen it?'

'Yes, I'm watching it now.'

'Eric, there's about a hundred of us here at the airport and we've chartered a flight. We're relying on your word. Are *you* 100 per cent sure it's a Brown Shrike?'

'Yes, come on over,' I replied.

It is right that birders should expect the person who runs the bird information line to know what they are talking about. The above is perhaps one of the more extreme examples of that kind of pressure. There was also the political side to running the BINS line. On occasion, it often took a great deal of delicate discussion to get to the truth of some bird news. This might occur when, for example, I got a call from an unknown birder who was reporting a bird that was totally out of season or that was really 'off the wall'. I would contact the person to determine if the bird was what they thought it was. Often, it would be quickly established that a mistake had been made and it was important for me to deal with such situations with the diplomacy of a veteran negotiator with the UN Security Council.

With the BINS line now the key source of bird news in Ireland, I had built up a network of contacts all over Ireland and Britain. These were birders who kept in touch with me, and I, in turn, kept in touch with them. It was a two-way thing. Whenever I received news of any major bird, as well as updating the BINS line I also made lots of calls to get the news out onto the grapevine. When texting became commonplace, I would text over 100 birders with a 'red alert' text message. So, almost every birder would hear about mega birds without having to ring the BINS line at all.

Despite the success of BINS as a very efficient form of gathering and communicating Ireland's bird news, it was only a couple of years from the launch of the BINS line before the first rumblings of discontent began among a very small number within the birding fraternity.

'It's wrong that we have to pay to hear bird news.' This was the common complaint.

When I was updating, I spoke very fast and gave the major news first. Many times, you'd hear what you wanted to hear in the first few seconds and could hang up. So, for example, the update might start with:

'8.20 a.m., Monday, 29 Sept. The Canada Warbler in Clare has now moved to the garden opposite the pub at Kilbaha this morning …'

This vital bit of news was exactly what you needed to know as you arrived into Kilbaha, having driven overnight from Dublin. It would have cost you less than 10p and would save you hours of anxious searching for the bird in the wrong location.

One story actually captures this slightly obsessive determination to avoid having to pay for bird news.

When Ireland's first Mourning Dove was found on Inishbofin island, Co. Galway, in 2007, Anthony (who found the bird) rang me with the news and details. The BINS line was updated with a red alert and almost every birder in Ireland received a red-alert text from me.

One birder was unable to twitch the bird for several days due to other commitments. The Mourning Dove was moving around the island a lot during its stay and often took an hour or two to locate each morning. Eventually this birder was free to twitch the dove.

He drove to Dublin Airport and parked his car there for the day. He then got a flight from Dublin to Galway. At Galway Airport, he hired a car and drove to Cleggan. There he had arranged in advance to charter a fast boat for the day (the ferry was not leaving and returning at the right times). He arrived on the island and began looking for the Mourning Dove.

Another birder met him. He had been walking around like a rabbit in the headlights for over an hour and was frantically looking for the dove.

'Have you not seen it yet?' this birder asked him.

He shook his head.

'What are you doing down here?' the other birder said. 'It's up by the church! Sure the precise location of the dove has been up on the BINS line this past hour. That's how I found it so easily.'

'Well, I didn't ring the line!' came his reply. 'I'm damned if I have to spend money to find out where the bird is!'

He then ran off towards the church, where he saw his bird.

After that, he got his chartered boat back to Cleggan, brought his hired car back to Galway and caught his flight back to Dublin. At Dublin Airport, he paid his parking fees and drove home. The entire twitch probably cost him €300 or more. Yet, he risked not seeing the Mourning Dove because he did not want to take on the cost of ringing the BINS line. If he'd phoned the line that morning, he would have spent no more than 15c to get all of the details he needed to quickly achieve his aims.

This remains one of the most startling examples of illogical thinking I have come across. It's also worth remembering that this guy had heard about the Mourning Dove in the first place by getting a red-alert text from me.

It seemed that this small group of birders were reluctant to accept the BINS line as *the* central source for all the bird news. However, without the BINS line, if a birder needed to find out all of the news from all over country, they would have needed to make a lot of phone calls to 'grapevine birders' in every corner of Ireland. This would have cost a lot more than making one call to the BINS line.

They failed to accept that I had worked hard to create that huge network of contacts. British birders, visiting Ireland, now had a number to phone to report their findings. In the past, Irish birders would often only learn what visiting birders had seen weeks or even months later.

They failed to accept the BINS line as being on top of the latest news. When you were arriving on a headland to look for a rarity, BINS had the latest update on the whereabouts of the bird, what it was doing, when it was last seen, etc. News on so many rare birds might not have been released as quickly were it not for the BINS line. Every birder had the BINS line to thank for many ticks.

Ultimately, this very small group of people did not take into account the time and commitment that was required to maintain the BINS updates. In busy periods, the line might be updated ten, fifteen or twenty times every day. In some months, there might be 500 or more updates made. Then, every evening at 9.30, the full summary of all the day's news would be read out in order of rarity and with full details. My life was tied to bird news and reporting on bird news. I was tied to doing that 9.30 p.m. update no matter what. It curtailed things like going to the cinema or theatre. If I was having a family meal, I would need to excuse myself to do my updates. When I went on holidays, friends such as Victor Cashera and Paul Kelly managed the BINS line for me.

I hated the politics involved in running the line. However, what I hated most was the negativity I encountered from a small minority of birders (and they were a very small, albeit influential, minority). I very quickly learned how much safer it might be to do nothing in life. Doing something new or different always runs the risk of drawing negativity from one quarter or another. I am happy to say that I also quickly learned to ignore such negativity and to enjoy all the positives that running the BINS line brought. It is a lesson I have carried with me in life.

I am proud to say that BINS provided a top-class bird news service. On some days during the peak autumn season, the BINS line received 300 calls or more from birders wishing to access the latest news. After twenty years of success, the arrival of texting, the internet and, a little later on, the age of tweeting and Twitter, made me realise that the BINS line had run its course.

In December 2009, I announced that the BINS line would cease. The last update took place on Wednesday, 9 December 2009. Hundreds of birders rang up to listen to my final 9.30 p.m. BINS update. It was the last of over 7,000 consecutive 9.30 p.m. BINS updates. There were just two occasions when I could not do the evening update. The first was when there was a technical fault in the network in Dublin. The second was when the phone lines went down on Cape Clear and I could not make or take calls (before the days of mobile phones). For months after I closed the line, I still found myself getting up from whatever I was doing at 9.30 p.m. to do my update.

I realise that, during these twenty years, birding and bird news were my life (the understatement of the year). As for the BINS line, I loved it and hated it all at the same time.

The politics and negativity got me down at times, but I loved the buzz and excitement around the finding and reporting of rare birds. I loved the adrenalin rush that came with the knowledge that I was about to make a red-alert BINS update. I loved being in touch with so many birders across Ireland and Britain and the many friends I made. I loved the support I received from the vast majority of Irish birders.

These were exciting and changing times for Irish birding and I enjoyed being at the forefront of it all.

Soon after that last BINS update, many lovely tributes were paid to me in personal phone calls and texts, and publicly in emails and messages posted on websites. It really was touching to see that so many birders appreciated my contribution to Irish birding over those years. Some of the tributes were so nice that they prompted one person to question whether they were in fact obituaries. I assured him that rumours of my death had been greatly exaggerated.

I was very much alive and well.

Chapter 22

WORK IS THE CURSE OF THE TWITCHING CLASSES

The great Oscar Wilde once declared that 'work is the curse of the drinking classes.' The same can be said of the twitching classes, especially those among the birding family who are hard-core twitchers. True, having a job or, as some like to call it, a 'career', does pay for that good pair of bins around your neck and that scope you carry. It also provides money to buy a half-decent car that won't conk out on you half way to Cork at some godforsaken hour during a storm in October. It also pays for the petrol that gets you to that headland. Of course, a job helps feed the kids if you have them, but, for the hard-core twitcher, that fact is the least important reason for having a job. But the truth is that, while work does pay for you to indulge in your twitching, it can sometimes get in the way of twitching. I know this all too well, but an example of this terrible conflict that stands out is an experience I had in September 1991.

Saturday, 21 September 1991 saw me in Co. Wexford, leading a Birdwatch Ireland Tolka Branch outing around the hotspots of Tacumshin and Carnsore Point. It was peak migration time but strong winds and heavy, prolonged rain resulted in nothing of note being seen. We had high hopes of finding an American wader, but despite our best efforts there wasn't even a Pectoral Sandpiper to be found. Three days earlier, news broke that an Upland Sandpiper had been seen on Dursey Island off south-west Co. Cork. For those who don't know, Upland Sandpiper was (and still is) a very rare vagrant from the prairies of northern Canada. They are odd-looking birds, with an alert upright stance, a large eye, a long tail and a shortish bill. This was only the tenth time the species had been seen since records began in 1855. This really was a rare bird.

Many birders headed to Dursey on 21 September and successfully twitched the bird. I was committed to leading the group around Wexford and could only wish that I was with them on Dursey. However, when I was finished with the group, I headed off to Cork with the aim of being in the right place to get to Dursey on Sunday if the bird was still around.

I stayed in Cork overnight and waited for news from Dursey on Sunday morning. It was negative: the Upland Sandpiper wasn't to be found. It had probably left overnight. However, a North American warbler called a Red-eyed Vireo had been found at Crookhaven and it was showing well. That was until I arrived at the site, whereupon it duly disappeared into the vegetation never to be seen again. It was turning into one of the worst birding weekends for years.

By the end of the day we had seen nothing and we turned back for the long road to Dublin. It's always a long, exhausting trudge back home if you've missed the bird you had travelled to see. The drive got even worse when I checked the BINS answering machine. The news that it delivered sent me into the depths of depression. First, birders who had visited Dursey late in the afternoon had relocated the Upland Sandpiper. Worse still, a Lesser Grey Shrike had been found in Dungarvan in Co. Waterford. This mini bird of prey is a European species and this was only the fourth time the species had been recorded in Ireland. Despair descended on me. It was Sunday night. We had stopped for grease at a traditional greaser in Urlingford, Co. Kilkenny. It would have been easy to turn the car around and head to either Waterford or back to Cork, except for one thing – work.

I had to get to work in the morning. By this time, I was working for Eircom, it having emerged when the Department of Posts and Telegraphs was split in two: An Post and Telecom Éireann (later Eircom). I had reached the dizzy heights of being a manager dealing with corporate business customers. The role demanded some level of responsibility which was not always welcomed. I had been putting in the groundwork for a major business meeting scheduled for Monday afternoon, and, shrike or no shrike, I had to turn up in suit and tie for that meeting or there would be hell to pay. I had no choice. I had to go to work.

As I drove home, I hatched a plan. I would go into work in the morning and brief my manager, a lady with whom I did not get on at all (to put it mildly). I would also have my birding gear in the car. In the meantime, I would check on whether the shrike and sandpiper were still present, and if they were I would then take some annual leave. If it all went smoothly, it would be a perfect plan.

Monday, 23 September dawned bright and sunny. I was up and into work early, making sure that I parked the car in such a position that I could make a quick escape by lunchtime if the news was positive from either Waterford or Cork. I was at my desk and working on a briefing document for the meeting. I was meant to be going to the meeting that afternoon, but I figured that if I had a clear brief prepared I should be able to convince my boss that I was not needed. I anxiously awaited news.

9.05 a.m. and I got a call to tell me that the shrike was still present and showing well. Operation Shrike was put into action.

I quickly rang Mick and told him that the twitch was on.

'Get into my place for about midday,' I instructed him. 'Hang around reception and I should be ready to go at about 12.30.'

9.40 a.m. and 'she' arrived into the office. I asked to speak with her but she quickly informed me that she was late for a meeting and would speak with me when she got back. That was okay. It gave me more time to make plans and more phone calls to establish whether the Upland had been seen.

10.30 a.m. and I got the news that I was hoping for: the Upland Sandpiper was also still present, albeit a little elusive and feeding in cover towards the very tip of Dursey Island. After such a dreadful weekend, I had a great feeling that these days would more than make up for it.

In the meantime, Mick was on the bus from his home in Blackrock and heading into the city centre to meet me in my Eircom office building.

11.50 a.m. and 'she' returned. She went into her office and closed the door. She had a real sulky head on her, more sulky than usual (or perhaps I just imagined that). I knocked at her door and she 'permitted' me to enter.

I sat down and suggested that we speak about the important meeting that afternoon. I handed her the briefing document I had prepared and suggested that I talk her through it. I said that there was probably no real need for me to attend the meeting. She glanced at the brief and nodded in agreement. So far, so good … Operation Shrike was going to plan.

Before talking her through the document, I felt that I needed to approach the rather delicate matter of needing to take the afternoon and the following day off at short notice.

'Em, before going through this, I would like to ask if it's okay for me to take this afternoon and tomorrow off.'

She looked at me, her eyes cutting me in half. 'No, it's not okay,' she replied sharply.

I had anticipated that this might be the result of the first salvo, so, undeterred, I continued. 'I know it's short notice, but it's very important for me to take this time off,' I said.

This was the truth. It was very important for my Irish list as well as my emotional and mental health.

'I said *no!*' she snapped.

This was not going to plan.

'I really do need to take this time off,' I said again.

'Why?' she demanded.

I called on my inner strength to keep calm and focused. By now I realised that the truth was the last thing she needed to hear, so my years of experience took over.

'I need to attend a removal this afternoon and a funeral in the morning,' I found myself saying.

'Well, you can take an hour off this afternoon to go to the removal and an hour in the morning to go to the funeral,' she said.

Feck it, she was good. I felt a sweat building on my brow. This was going to take some effort, but the thought of missing those birds drove me on to new, higher levels in this deadly game.

'That doesn't suit, I'm afraid. The funeral is in Limerick.'

To this day I have no idea why I chose Limerick, but there it was.

She looked at me. We were like two gunfighters in the Wild West sizing each other up. She had the advantage: she was the boss. After what seemed like an hour, she shook her head.

'No, that's out of the question. You can't have time off,' was her verdict. 'Now, let's get on with this briefing document.'

It seemed like game, set and match to her. However, from my inner core, I produced an ace card. It was the final play and it was all or nothing. I put my sad face on (birders have sad faces when they fear they might miss rare birds). Then I played my ace.

'Okay, that's fine,' I said. 'I fully understand, but before we go through this document, can I just take a few minutes to make a quick phone call to my mother?'

This caught her off guard.

'Your mother?' she asked.

I had her on the hook and by God I was not going to stop now.

'Yes, my mother,' I continued. 'You see, I promised that I would take my mother to the funeral. Eh, it's her brother who died suddenly over the weekend. She is the youngest in the family and he was the only sibling she had left.'

It was so good that I almost believed it myself (for the record, my mother never had a brother in Limerick). In addition, my mother's brother was an ideal candidate for a funeral as my manager would not have known my mother's maiden name. Even if she checked the death notices in the paper, she would be none the wiser.

Before she could say anything else, I kept going. I was on a roll and she was on the ropes.

'I need to let her know that she can't go … and there is no one else who can take her. I really need to let her know that she can't make it to her brother's funeral …'

I let the sentence fade for effect. It worked perfectly. She looked shaken. She sighed deeply before uttering the words I most wanted to hear.

'Well, if that is the case, then, yes, you can have the time off.'

With that, I went in for the double bluff.

'Only if you're sure it's okay,' I replied. 'I mean, I'm sure my mother will understand if I can't bring her.'

'No, no, ring her right away and tell her you can bring her,' she insisted.

'Okay, thanks … I am sure she will be very thankful to hear that,' I said quietly.

I think she heard the emotion in my voice. There *was* emotion in my voice. The thought of missing two ticks is enough to bring any birder to tears.

By now, Mick had arrived at reception. He was sitting, cool as you like, telling the security man all about the Lesser Grey Shrike and Upland Sandpiper, blissfully unaware of what was unfolding on the first floor.

I asked to be excused for a few moments on the pretence of calling my mother to confirm that she was able to go to her fictitious brother's funeral, and I left her office.

I raced down to reception and found Mick all ready and waiting, a scope and battered tripod in one hand, bins around his neck and a small bag over his shoulder. He always travelled light. He was in his usual birding attire of big, baggy black jumper (with numerous holes), paint-splattered faded jeans with an enormous hole in the right knee and a dirty green waterproof jacket that had seen better days. He wore dirty hiking boots, which always seemed a size

too big for him and proudly displayed a selection of mud from
many esteemed birding areas such as Ballycotton and Tacumshin.
He stood out like a sore thumb, causing an alarming untidiness
on such posh corporate reception furniture, upon which also sat
another individual who was dressed in an immaculate suit.

'Get the fuck out of here!' I said quietly to him. 'Seriously, Mick,
get the fuck out of here as fast as you can ... I'll explain when I see
you. Make your way to Newlands Cross and I'll pick you up at the
bus stop.'

Michael knew by the tone in my voice and the wild look in my
eyes that things were very serious. He left the building immediately.

I went back upstairs and, taking a deep breath, went back into
the office to brief my manager on the meeting, but not before
thanking her on behalf of my mother for being so kind and
understanding.

By 12.40 p.m. I was on the road. I picked up Mick on the Naas
Road and we laughed all the way down to Dungarvan after I
recounted what I had to go through to get time off.

Arriving in Dungarvan, we followed directions to an area on
the outskirts of the town and were relieved to find lots of birders
watching the Lesser Grey Shrike. It was a beauty, with a lovely grey
back, black eye mask and black wings. It was showing very well.
It was causing quite a stir and lots of people were arriving just
to see what the fuss was all about. Among them was a reporter
from the local radio station who wanted to know how so many
birdwatchers had heard about the shrike so quickly. As I was the
person running the BINS line, and therefore responsible for so
many people learning about the bird, the reporter was pointed
in my direction. However, I became very media shy, refusing to
speak with anyone. I had visions of someone from Eircom in
Waterford hearing me on the radio and telling my boss about
it. The reporter found it very odd that not one person who was
watching the shrike was willing to speak with her. Of course, what
she didn't know was that everyone there was most likely skiving
off work; sick days, funerals, you name it – every excuse had been

delivered to a whole selection of bosses by just about everyone there.

With the shrike under our belts, we headed west and spent the night in Allihies, Co. Cork. There we were joined by three other Dublin birders who were also skiving off work. They had a perfect excuse. They had just returned from holidays and had driven straight from Dublin Airport down to Cork. They would tell their respective bosses that they had missed their connecting flight. This allowed them to save the funeral line for another twitch.

Dursey Island is hilly and windswept. The Upland Sandpiper had been present on hillsides covered by a dense growth of ferns towards the tip of the island. It was a long walk out, but we eventually reached the last site it had been reported from. We spread out across the hillside. It was dangerous underfoot and I had visions of slipping and breaking my leg. I could see the headlines in the newspapers and the report on RTÉ *News* about a birdwatcher being airlifted off Dursey Island. I could see 'herself' sitting in front of the TV watching pictures of me being taken off the island by helicopter. That would take some explaining.

Such thoughts were interrupted by a shout. The bird had been located. Within minutes, this trans-Atlantic vagrant was showing distantly. We had got our bird. There were handshakes and smiles all round. The mental and emotional strain of the previous day had been worth it all.

It's funny how the drive home to Dublin passes so quickly and easily when you see the bird you have travelled to see.

I returned to work the following morning and found it hard not to keep smiling to myself. Birders have such happy faces when they see the birds they hope to see.

I must end by saying that, in order to restore good karma to my birding, I informed Ma about the web of lies I had told in order to see the shrike and sandpiper. She laughed and rolled her eyes to heaven.

'Some day you'll be found out,' she warned.

I never was.

Chapter 23 ✈

DELIVERANCE

On 2 November 1996, Mick and I boarded a British Airways flight from Heathrow to Hong Kong at 9.50 p.m. It was the beginning of our four-week Australian birding odyssey and our minds were filled with thoughts of the adventure ahead. In all, it would take thirty-six hours to get from Ireland to Australia. In comparison to the long sea journeys the Irish of the past had to endure to reach Australia, this was nothing. It really hits home just how small the world has become in this age of jet travel.

Our stay in Hong Kong was a short one of only a few hours. We processed our tickets through the stifling heat and chaos of the transit desk before boarding a Qantas flight to Cairns.

Arriving in any new country is always a birder's delight. Arriving in a new continent is even better. Not only are there new birds to see, but more often than not there are also whole new families of birds to familiarise yourself with. However, we had done months of research and were armed with *The Field Guide to the Birds of Australia* by Simpson and Day, and a signed copy (hot off the press) of *The Complete Guide to Finding the Birds of Australia* by Richard and Sarah Thomas. These would be our bibles for the next four weeks.

We landed at Cairns Airport at dawn, got our car and set off on our birding odyssey. Mick was the designated navigator. I was the driver. From the airport, we drove to the Esplanade, a walkway overlooking the vast mudflats of Cairns. It took us over an hour to complete the 3 km journey, as our birding senses were bombarded with many new and colourful species. My first new species was a White-breasted Woodswallow. I always consider it a good omen if a species of swallow is the first new bird I see on holiday.

At the Esplanade, we had planned a rendezvous with an old Irish birding friend, John Grant, who now worked in the rainforests of Cairns. John had left Ireland many years earlier. It was 7 a.m. when we arrived and we were already in birding heaven. I won't go into all of the birds we saw, but it seemed that everything that moved was new to us. There were Mistletoebirds, Royal Spoonbills, Helmeted Friarbirds, Varied Honeyeaters, Spice Finches, Rainbow Bee-eaters and Rainbow Lorikeets. For wader fanatics like us, there were hundreds of Red-necked Stints, Broad-billed, Terek and Sharp-tailed Sandpipers, as well as Grey-tailed Tattlers, Eastern Curlews and Great Knots. It was almost an overload of birds. Before we even met up with John, I had seen twenty-one new species of bird.

It was great to be with an old birding friend. It had been a long time since John and I had met. In fact, the last time was a night spent looking at slides in Ronan Hurley's house when John was visiting Ireland some twelve years earlier. With John guiding us, we were in good hands. He took us to some of the other great birding spots around Cairns, including the Botanical Gardens. It was here that I achieved a landmark in my birding life: my 1,000th bird on my life list. John brought us to a large tree along a forest path, stopped and pointed up. There, almost invisible due to its cryptically patterned plumage, sat a very still and upright Papuan Frogmouth, a large, nightjar-like species. It looked like the stump of a broken branch. I could not have asked for a more dramatic and beautiful species for my 1,000th species of bird.

That evening, we followed John into the depths of the Warrawee Rainforest to the research station where he was working. The start of our trip was perfectly timed as we had arrived during a short holiday period when all of the students were away. We had the whole rainforest and the station to ourselves. The rainforest was set aside for study and was effectively off-limits. We could not have hoped for better. As dusk descended, I watched the ghostly shapes of Spectacled Flying Foxes (fruit bats) as they left their roosts to feed in the forest. By the end of our first day, we were exhausted. But I had seen sixty-six new species of bird. This was by far the largest number of new species I had ever seen in one day. I drifted off to sleep to the deafening sounds of insects, frogs and the whistling calls of Lesser Spotted Owls that sound like dropping bombs. It was magical.

The dawn chorus of a rainforest is an experience of a lifetime and at Warrawee it was a dawn chorus like no other. It almost matched (but didn't beat) that very first dawn chorus I had with Da in Finglas all those years ago. This rich rainforest was brimming with birdlife. By 5.15 a.m., the calls of Tooth-billed Bowerbirds, the explosive calls of Whipbirds and the calls of God only knows what else echoed around the station. It sounded like the soundtrack from old Tarzan films. By 5.20 a.m., we were up and birding. It was the start of a long and bird-filled day. I fondly remembered that first Finglas dawn chorus even more when, as we walked along a narrow track, we came upon a large, thrush-sized bird feeding in the upper canopy of the trees. We trained our scopes on it. It was a velvety purple colour in the sunlight (black in the shadows), with a shimmering, iridescent greenish-purple throat and greenish underparts. There in all its glory was a male Victoria's Riflebird. My mind was reeling with memories of *Purnell's Illustrated Encyclopaedia of Animal Life*. This bird was in Volume 1 on page 202. I stood in the humidity of this rainforest, savouring my first (and long-waited) moment with a *genuine* bird-of-paradise. It was the most exciting and intensive birding I had ever experienced. We encountered so many species that I will not even attempt to recall them here.

Following several wonderful bird-filled days, we said our good-byes to John and departed Warrawee, heading inland towards the desert regions of the Outback. Our journey took us along dusty roads towards the small village of Herberton, a stopping-off point before heading out into the most arid regions of Queensland. Driving along the roads of this remote region was a very new experience. These are narrow tarmacadam roads just wide enough for one car; when another car comes in the opposite direction, you both go off the road onto the rough surface along the sides. This routine took a little getting used to, but the long straight roads made birding and driving easy. It was also strange to see hundreds of pink parrots (Galahs) flying off the side of the road as the car approached, in the same manner as Rooks do in Ireland.

As we approached Herberton, the feeling of true remoteness was overwhelming. There were no houses and we met fewer and fewer cars along the road. As we drove, we took random turn-offs here and there to see what birds we might find. There were hundreds of such turn-offs and each was pretty much the same. One turn-off we took seemed to go on for ever, but we were in no hurry and we followed our noses. After a dusty 10 km, this small dirt track came to a T-junction. We flicked a coin, which decided that we go left. This dirt track continued on for another 5 km until we eventually came to a dead end by a small stream. It was extremely hot and dusty, so any water might be good for attracting birds. We were a long, long way from civilisation, so it felt like a prime place to bird. We parked the car under the shade of a tree near a gate and began walking towards the stream. We had not taken more than three steps from the car when we heard a shout.

'Hi, matey! Whatya doin' here?'

We looked around to see a man in shorts and a T-shirt coming out the gate. In his right hand was a meat cleaver. He looked like someone from a cult horror movie and, if I'm honest, I got the sudden feeling that we might be next on his victim list.

'Eh, whatya doin' here, mates?' he asked again.

I really didn't like the look of this. We were in the middle of nowhere and were being challenged by a guy with a meat cleaver. I couldn't help but notice that there appeared to be bloodstains along the length of its cutting edge.

'Eh,' I replied, 'we're birding.'

I didn't expect that he would even understand my reply. I felt a slight shiver of fear run down my spine. I tensed up, expecting to have to defend myself. I noticed that Mick was also standing tall and had taken a slightly defensive stance.

He looked at us. He could see the bins around our necks.

'You're birding, eh?' he said slowly. 'So, have ya ever fed a 'burra?'

It was an unexpected question, to say the very least. Kookaburras are very large kingfishers that eat everything from insects to snakes, but they are rarely found along rivers like our Kingfisher. We had seen lots of them perched on the telegraph wires on our journey, but, no, we had never fed one. We shook our heads.

He looked at us again. All I could see was this bloodied meat cleaver in his hand.

'Where are you guys from?' he asked.

'We're Irish,' I explained, hoping that this fact might somehow save us from being butchered on this dusty outback road in the middle of nowhere.

With that, he smiled and rushed forward to vigorously shake our hands.

'I don't believe it!' he said. 'You're Irish! I'm Con Leahy. My great grandfather came from Cork. It's great to meet some Irish guys. Come on in. I'm just cuttin' up some 'roo [kangaroo] I picked up on the road and feeding the 'burras.'

He opened the gate and invited us in. Mick and I looked at each other. Were we mad to even think about going in? I looked at this guy's bloodied meat cleaver. I could almost hear a banjo picking out those famous notes from the film *Deliverance*. However, nothing ventured, nothing gained seemed to be our mutual feeling, so we followed in through the gate. I left the gate open behind me and the car unlocked in case we needed to make a quick escape.

The small house he lived in was about 500 metres down a dirt track and close to the river. As we came to the house, we saw he had a tray of freshly cut meat outside (the last of the people who came by here before us, I contemplated silently). He picked up a piece of meat and held it out on his hand. He looked around him.

'Kooookkkkoooooo!' he shouted, frightening the life out of us.

Within seconds, the distinctive laughing, churring calls of a Kookaburra called back and almost instantly a bird perched in one of the trees above us. It threw its head back and called loudly right over us. Con smiled at us.

'Kooooooooookkkooooo!' he repeated, shaking the piece of meat so that the bird could clearly see he had food.

With that, the bird flew down from its perch, landed on his wrist and took the meat from his hand. It was a sight to behold.

'Wanna have a go?' he asked us.

We didn't need to be asked twice. We took it in turns to hold out our hands as Con called several Kookaburras in to feed from them. I have to admit, it was one of the most surreal episodes in my life. Here I was, literally in the middle of nowhere, feeding a Kookaburra by hand, while a cleaver-wielding stranger looked on.

When we had used up all the meat, Con invited us in for a drink. We declined. I really had seen too many of those cult horror films. It would be the perfect way to get rid of us – to feed us to a hungry mob of kingfishers. No, going inside for a drink was not a good idea. We thanked him for his hospitality and for allowing us to share such a unique experience with the Kookaburras, and said our farewells.

'Have ya ever seen a Platypus?' Con asked as we were turning to go. 'If ya haven't … well … I can show ya where the critters can be easily seen around here.'

This offer stopped me in my tracks. My mind reeled with my childhood dream of one day seeing a Duck-billed Platypus, one of only two species of egg-laying mammals in the world. Again, I could almost see the page open in my *Purnell's Illustrated Encyclopedia of Animal Life*. I knew they were hard to find and I had no expectation of seeing one on this trip.

He hasn't tried to kill us yet, I thought to myself.

'Yes, we'd love to see a Platypus,' I replied, ignoring the banjo tune that had started up again in my head.

Con was delighted. We got the impression that he rarely saw people, never mind had visitors. Let's face it, he lived in such a remote place that I think the chances of anyone stopping by his gate were a million to one. He seemed determined to keep us around for as long as he could.

'Great, follow me,' he said, leading the way.

He led us past his house and down a narrow grassy track to the riverbank. He sat down and indicated for us to sit too. The river was clear and clean. We had been sitting for no more than five minutes before we heard a splash just down river. Con indicated that we watch the lower section of the river. Then up popped the most unusual head I have ever seen. It was beaver-like in shape and size but with a thick, duck's beak stuck onto its face. It was a Duck-billed Platypus. We watched, mesmerised, as this plump, brown, flat-tailed, web-footed curiosity of a mammal swam around the clear, still waters of the river. We could see it forage for food along the river bed. It was like watching a wildlife documentary. It was spellbinding. As quickly as it appeared, it disappeared again, and, in so doing, allowed us an opportunity to make good our escape.

We shook hands with our new 'Irish' friend and wished him the very best. Driving away, we were relieved that the encounter had been such an unexpectedly fruitful and fun experience. We really had feared the worst when we saw the meat cleaver.

Some five years later, I was giving a presentation on birds to a very welcoming audience in Sligo. There, I met Don Cotton, an old friend from my early birding days in Dublin.

During the course of the conversation, he told me that he and his family had been birdwatching and hiking in Australia during the summer just gone. We shared our respective highlights from our experiences there.

'Oh,' he added, as he turned to go, 'I met a friend of yours down there and he asked me to say hello when I saw you.'

'And who was that?' I asked, expecting that it would most likely be John Grant, as Don also knew John from when he'd lived in Dublin.

'Let me think … what was his name?' he said. After a few moments, it suddenly came to him. 'Ah … Con Leahy. That's who it was. Yes, Con Leahy!'

'Con Leahy?' I asked in total disbelief. 'How in the name of God did you meet Con Leahy?

'Well,' Don said, 'we were travelling out to Herberton and we took a random turn-off and followed our noses until we came to a river. Then this guy came out carrying a meat cleaver …'

It seemed impossible to comprehend but Don and his family had taken the exact same random turn-off among hundreds along the road. They eventually came to halt by the river only to be greeted by good old cleaver-wielding Con Leahy. It is hard to describe how remote and isolated this particular house is, yet somehow Don had encountered this same man at this same place.

Can you imagine Don's surprise when this stranger with a meat cleaver, on realising he and his family were from Ireland, said …

'So you're from Ireland, eh? Do you know Eric Dempsey?'

It really is a very small world indeed.

Chapter 24

A BIRD WORTH DYING FOR

T he pursuit of birds can bring you to remote and wonderful places on this planet. However, while I thought that my experience with good old Con Leahy in Australia was one of remoteness, it was nothing to what I experienced some years later on the roof of the world. It was 4 July 2001 and we were birding in the woodlands of Yaoziquo Forest Park. It was challenging. It was our first real test of birding at high altitude. We had landed in Qinghai Provence that morning and had commenced our journey as soon as we had touched down at Xining Airport. I was travelling with two other Irish birders (and photographers) and this was the first part of our journey across the 'the roof of the world', the Tibetan Plateau.

The Tibetan Plateau is an immense mountainous region, some 3,500 km by 1,500 km in size, with an average elevation across the entire area of more than 5,000 m (16,000 ft). It is a remote and wild land, with small settlements and towns scattered all across it, separated from each other by hundreds of kilometres. Remote and isolated Buddhist monasteries are nestled in deep valleys and on mountaintops. Tibetan Buddhism is that of the Dalai Lama. Tibet was an ancient theocracy that was toppled in the 1950s when China annexed the country.

Tibet is a land of high plains and mountain passes. Its southern rim contains Mount Everest and thirteen other peaks higher than 8,000 m (26,000 ft). Its remoteness and inaccessibility sheltered Tibet from the outside world for thousands of years. The Tibetan Plateau is not just the largest area at the highest altitude in the world today: many scientists consider it the largest and highest in all of geological history. It really is the roof of the world. This is the region we were setting out to explore over the next four weeks.

We were met at the airport by Lou (Kyi Wi Lou), an experienced bird guide who had travelled extensively across Tibet. He had arranged the entire trip, including our all-important visas. We were travelling in an old Toyota Landcruiser driven by the smiling Mr Dung (that really was his name). The car had no air conditioning and was full to the brim with gear. Besides our luggage, scopes, cameras and tripods, we also had basics that included at least three big drums of diesel and over 100 bottles of water.

As we headed from Xining Airport, the landscape changed from the arid semi-desert plains to almost Alpine-like mountains. The desert plains are at 2,400 m (7,800 ft), so we really had no sense of being at a high altitude. As we began to climb slightly higher, I felt it was a little difficult to get a good breath, but it really was nothing too drastic. As we climbed into the Yaoziquo Forest Park, we reached over 3,400 m (11,000 ft) and the air became even more rarefied. We knew that driving across the plateau meant that we would be constantly at heights of 4,000 m (13,000 ft) with the road crossing over even higher mountain passes as we went further out. There was only one road in and one road out.

It was one thing to sit in the car as we drove to this height; it was another to start walking and birding. Yet, the birds that were on show were more than enough to keep me going. A flash of pink wings on a greyish bird flying along the cliffs brought me my first tantalising views of a Wallcreeper. Colourful gems like Przevalski's and White-throated Redstarts posed for my approval, while Crested

Tit-warblers flitted in the trees. Arctic Warblers sang throughout the woodland, competing with Yellow-streaked Warblers. A Long-tailed Rosefinch showed itself briefly, while a large Kessler's Thrush fed like a Mistle Thrush at the forest edge. It was superb birding. I did not mind the powerful sun that penetrated my UV sunscreen, nor did I mind the feeling of a lack of air to breathe as I struggled to climb the steep paths through the woodlands. I was in Asian birding heaven.

We stayed in the Qinghai Hotel in Xining City that night. It was a comfortable hotel and allowed us to relax and prepare for the next couple of weeks. Tomorrow, we were setting off into Xinhai County. It would be an adventure of a lifetime.

The following day, leaving Xining behind us, we drove through the spectacular scenery of high mountains and steppe habitats. We climbed around the Hungyuan Gorge, with its deep river valley below, before passing out onto the Tibetan Plateau. The roads were good as we entered the vastness of the plains.

It was here that we first encountered the famous yak herders of Tibet. Yaks are little hairy cow-like animals. One small encampment of yak herders was set in the most pristine valley, with lush green grass and clear ice-cold fresh rivers and surrounded by snow-capped mountains. The tents were wigwam-like with colourful bunting flapping in the wind. The yak herder society is unusual in that it is female-dominated. One woman has several husbands, who are usually from the same family. In this way, a family can keep their yak herd intact. Yaks are everything to these people. They drink their milk, eat their meat, and use their wool for clothes and their skins for tents. They even use their dried dung as fuel. It was unreal to see these herders gallop across the plains on their short, sturdy ponies. It was like taking a colourful step back in time.

They offered us tea, which we very politely declined. Tibetan tea is not quite like a nice cup of Irish Barry's or Lyons' Tea. It is drunk with large dollops of yak butter and so becomes a floating, greasy mess. For an Irish tea drinker like me, Tibetan tea is a truly disgusting brew. I reckoned even Da would struggle to drink it.

It was also along these roads that we encountered our first Tibetan monks and pilgrims. These were walking to Lhasa, a journey that might take years to complete. With each step forward they took, they would bring their hands together, drop to their knees and lie prostrate on the road. They would then get up, take another step and repeat this whole process again. With over 2,000 km to go to Lhasa, this demonstrated an extraordinary level of faith and commitment.

The plateau at this point was at 3,000 m (9,750 ft) and, while I found the air a little thin, again, I felt it was no real problem. We climbed up above the plains and crossed our first really high pass, the Heka Pass, at just under 4,000 m (13,000 ft). This was our first real Tibetan pass and it was adorned with prayer flags, each sending a wish to the skies from people who had passed through. These prayer flags were like small pieces of cloth strung along the pass. It looked untidy and dirty but this was the great tradition. At each pass we crossed, the prayer flags flew. People placed stones in little piles and threw prayers to the winds on small pieces of paper that floated high into the mountains.

It was here that I got my first taster of what altitude sickness might be like. My breath was laboured and I felt a slight dizziness from the lack of oxygen. However, the feeling soon passed. It was a world of rosefinches and snowfinches, of larks and enchantingly tame birds called Hume's Groundpeckers, which fed like Starlings in small parties on the short grass. From the Heka Pass, the surrounding high mountains were snow-covered and the Himalayas stretched for as far as the eye could see. Spectacular is a word often used to describe scenes like this, but that word does not come close to capturing the views that greeted me that day. The immensity of the mountains and the vastness of the landscape are next to impossible to describe.

We stayed that night in a small hotel in Xinhai. We were at 3,400 m (11,000 ft), so this was good for getting acclimatised to the height. I slept well and set off at six on the morning of 6 July feeling refreshed and full of energy. Today we were targeting Tibetan

Snowcock, a large grouse-like species that inhabit the highest slopes. They are found in the mountains of Tibet and nowhere else in the world. It was one of two species that I really wanted to see on this trip. The other was a very strange species called Ibisbill. It is found along the fast-moving high rivers of the Himalayas and is like a cross between a Ringed Plover and an Ibis. Ibisbills are a family and species all to themselves; there are no other Ibisbills in the world.

As we travelled into the mountains, the road surface suddenly changed from a reasonable tarmacadam to that of a rough, dusty dirt road. We had left the first stages of the new road system being built across Tibet and had started along our first kilometre of dirt road. Within seconds, we were covered in dust and dirt. The road would be like this for the next 1,500 km.

Climbing high above the plateau, we eventually stopped at Erla Pass. Our arrival was met with the alarm calls of Black-lipped Pikas (small little rodents that have black lips) and the louder, barking calls of Marmots (another large rodent). These became familiar sounds over the following days and weeks. They were the sounds of the mountains. Along the road, we'd passed Upland Buzzards sitting on posts, while up here giant Himalayan Griffon Vultures soared on broad, long wings. I hopped out of the car to enjoy these vultures and immediately felt slightly dizzy. It was difficult to catch my breath. Erla was just below 4,500 m (14,500 ft), so it was not too surprising.

As we boiled up noodles in our billycan at the side of the road, I enjoyed watching my first Tibetan Larks and a Blanford's Snowfinch. Then, over the pass came my first Lammergeier Vulture. It was a beauty: an adult with that classic yellow head and long black beard of face feathers that gives them their other name, Bearded Vultures. These species specialise in eating bones to extract the marrow from the inside. They will bring a bone to a great height, drop it onto the rocks below and then swallow large hunks of broken bone. They have very strong digestive juices which can break down bone. It soared over us with long wings and showed off its long, wedge-shaped tail. I stood back, savouring this special moment. This was a truly amazing place.

As I ate my can of noodles, I looked out over the terrain we were about to cross. The high snow-capped mountain ahead rose gently above open grasslands and scree slopes. Tibetan Snowcocks are usually found along the snow lines, feeding on seeds and insects. They are pale brown with grey heads and white underparts. Looking at the slopes, I realised just how perfect such colours were for camouflage. There might have been a million Tibetan Snowcocks on the slopes but they would be next to impossible to see from here. We would need to get close to the snow line if we were to have a chance of seeing them. Even though it was early, the sun was already quite intense and it was hot. It would be a hard hike, but they were worth the effort.

I set off with two bottles of water in my bag on my back, my camera on one shoulder, my bins around my neck, and a scope and tripod on the other shoulder. I was a bit weighed down, but these were all important pieces of equipment to carry. I was not going to leave the camera behind for fear of missing out on any good shots. My scope was essential to scan the slopes as we neared the snow line. Off we set, me walking and birding up the mountain at a brisk enough pace as if I was hiking up Bray Head in Co. Wicklow. Horned Larks flew ahead of me. Hume's Groundpeckers watched me. On I went, enthused and excited by the birds I was seeing, determined to reach the snow line where a Tibetan Snowcock might be waiting for me.

'Tibetan Snowcock is worth this effort,' I told myself as I began to pant and gasp a little.

On we went, climbing to over the 5,000 m (16,000 ft) mark, according to the small altimeter I carried. I stopped and scanned the mountain from this point but could see no sign of the birds. So I decided to climb a little higher. By now, I was feeling slightly dizzy. I drank some water. That helped. I walked on and, upon reaching 5,300 m (17,000 ft), I stopped abruptly. I didn't stop because I saw a bird: I stopped because I suddenly felt very weak.

I sat down for a moment to catch my breath – but I couldn't. I was finding it difficult to breathe. Then I felt my neck aching. It was like a bad crick in my neck, like the sort you get when you've slept awkwardly. Then the pain slowly rose along my neck and crept along the base of my skull. It was a very strange feeling. Slowly it continued over the back of my skull until it had the whole of my head in its grasp. It felt like my whole head was in a vice. I gasped for breath. How I longed to just get a good breath of air!

Then, as if it was acting according to a plan, the pain decided to force its way down onto my shoulders. My shoulders ached. The pain slowly descended into my chest. My head was spinning. My legs went to jelly. I tried to drink water but threw up. I lay there for a few moments. In my enthusiasm I had climbed higher than the others. They were well below me. This feeling was like nothing I had ever experienced. If you can imagine the worst bout of seasickness, add intense pain to your whole body and head, throw in not being able to breathe and then multiply that by ten – that would only come close to how I felt on the mountain at that moment.

Lou was suddenly by my side. He instantly recognised that I was in big trouble. He helped me to slowly retreat back down the mountain. We eventually reached the car. I was in bits. We had not seen our Tibetan Snowcock and at that point I could not have cared less. This was as ill as I had ever felt. I didn't care if I never saw another bird again; I just wanted the pain, the dizziness and the nausea to go away.

The only cure for altitude sickness is to get down to a lower altitude. The problem was that we were on the Tibetan Plateau; there was no real lower altitude to get down to. We had the choice of going back, abandoning this trip of a lifetime, or continuing on in the hope that I would adjust to the altitude. There really was no choice: we would continue on.

We moved off the Erla Pass and headed out across the plateau. I was sitting like a corpse in the back seat trying to drink water. The car was hot but leaving the windows open meant getting covered in sand and dirt. On we went until we managed to drop down towards a water spring at around 3,500 m (11,300 ft). This small drop in altitude was enough to allow me to recover a little. I hoped that this would be the only serious bout of altitude illness I would suffer. However, as I mentioned earlier, driving across the plateau meant that we were constantly at heights of 4,000 m (13,000 ft) with the road crossing over higher mountain passes as we went further out. So off we set again and, as we again climbed higher, the full scale of the condition assaulted me once more. The road was bumpy and dirty. I just wanted to lie down somewhere and die.

I don't remember much about the following 150 km. Lou wanted to keep me down at moderately low altitudes so he decided we would stay in a dusty little village called Hua Shi Xie. We found a guesthouse and, while Lou went in to check it out, I remained in the car. Within minutes, the car was surrounded by over fifty people staring in at us. It was like being in a cage in a zoo. This staring became a regular feature of every little town, village and settlement we stopped in or passed through. In the street, people would just stop and stare at us, sometimes right into our faces. It was very disconcerting, but by the end of the trip it was something that I didn't even notice.

The guesthouse at Hua Shi Xie resembled a cattle shed. In fact, a cattle shed would seem like five-star accommodation in comparison to this guesthouse. Even Lou found it terrible. He had warned of 'primitive conditions' in his preparatory notes, but this was far below primitive (it certainly wouldn't get five-star ratings on TripAdvisor).

We decided to move on to Madoi, another busy, dusty, dirty village. Lou was unhappy that we had to stay here because, at 4,200 m (13,650 ft), it was still too high up for someone suffering from altitude sickness. However, the guesthouse was the best we could find and by this time all I wanted to do was to lie down. The guesthouse room was small, hot and damp, with a small stove in the corner burning dried yak dung. The smell of a yak-dung fire is hard to describe, but, surprisingly, it's not as unpleasant as you might expect.

The room was also filthy. The bedclothes were dirty. The pillow was stained. Black hairs were on the sheets. I didn't care. I unfurled my sleeping bag and lay down. My head was exploding. My stomach was churning. My chest was being squeezed. I could not breathe. I began to be sick again. There was no toilet for me to be sick into, only a large plastic basin. The basins were both for washing yourself and for toilet duties. I then had to bring the contents of my basin and pour them into a large communal bucket in the hallway, into which those in all the other rooms emptied the contents of their respective basins. The hallway where the communal bucket was kept was stinking. This was my home for the night.

I tried to sleep, but the nausea, the pains in my chest, the feeling of my head being in a vice, my aching neck and my lack of breath kept me awake. It was a long, long night. It was the first time in my life that I would have been quite happy to 'shuffle off this mortal coil'. It was the worst night of my life. At some stage I must have fallen asleep. I was surprised to wake up. That is an odd statement to make, but it best describes how I felt when I awoke. I didn't know I had fallen asleep but I was surprised that I had survived long enough to wake up.

That morning, Lou was more than relieved to see that I was still alive. I felt marginally better and managed to drink some water and eat a few biscuits. Despite Lou's concerns, I insisted that we continue on.

We left Madoi. I continued to feel very ill as we journeyed out across the plains and mountains. We crossed the Yellow River to begin our climb towards Qinghai Lake. I could barely stand to enjoy the wonderful sight of a Black-necked Crane, one of the world's rarest and most beautiful cranes. I was so ill that even looking through my scope made me feel dizzy and nauseous. At this stage I could hardly keep water down. I was weak and dehydrated.

On we went for over 250 km on the bumpiest, dustiest dirt road imaginable. At the prayer flags of Kana Shan Pass, I thought my head was about to explode. I simply could not take the neck pain and headaches. I could not breathe. We were at almost 5,500 m (almost 18,000 ft). The pain was excruciating, but thankfully we dropped very quickly back down to 4,000 m. And so we went, climbing passes and then dropping back down. We drove over the great Yangtze River, but I couldn't have given a 'yangtze' about whether I saw it or not. It was one of the worst drives of my life.

We were heading into a small town called Yushu. It was larger and slightly more modern than the other towns and villages of the region. It was also at the lower elevation of 3,700 m (12,000 ft). Lou hoped that we would find a good hotel to stay in, a place where I might get a chance to recover. Just outside Yushu, as we were driving along the fast-running river, we suddenly stopped.

'Ibisbill!' Lou said.

There below us, feeding slowly along the shallow parts of the river, was a pair of Ibisbill, one of the most sought-after birds of the Himalayas. I barely raised my bins to look at them. This was a bird that I had always wanted to see and I was feeling so ill that I had little or no interest in seeing it. That was how bad I was feeling.

Yushu was busy and hot, but Lou found a brand-new hotel that had just been built. We booked in there. It was bliss. It had a clean bed and sheets, a shower and a toilet. The simple things in life really are so good. Being at this slightly lower altitude, I felt I could catch my breath and the pain in my head and neck began easing slightly. I even managed to eat something light and drink three bottles of water. I was exhausted. While everyone else headed off for food, I went to bed. I slept like a log.

The morning of 8 July was bright and sunny. I awoke and took a deep breath. It came in clean and full. I opened my eyes. Slowly, I lifted my head off the pillow. While I still had a slight headache, the vice-like grip had gone. I turned my head. My neck was pain free. I took several deep breaths. It was as if the whole weight of altitude sickness had been lifted clean off my shoulders, chest and head. I had survived it.

That morning, we retraced our steps along the river and found our Ibisbill. In fact, there were five Ibisbill feeding along this stretch. They were superb birds. I took deep breaths as I sat just enjoying them. I really appreciated the fact that I could breathe easily. I appreciated this chance to share my time with these special birds. We sat by the river and I ate Tibetan bread and some fruit. It was good to eat again.

With my strength coming back, we decided to try to climb into one of the high valleys to see if I could survive it. We climbed to over 4,800 m (15,600 ft). I stood out and took a deep breath. The air was as pure and as oxygen-rich as I have ever breathed; I could have been standing on the North Bull Island looking out to sea. We ventured up into the Nipa Valley where we found magnificent species such as Godlewski's Bunting and Przevalski's Rosefinch (two rare Tibetan endemics).

In the Nipa Valley, I stared at a high, snow-capped mountain and thought, Hmmm … Tibetan Snowcock.

Armed with scope and bins, I walked out across the grasslands and started to climb the mountain at a steady and easy pace. It was here, as I took deep breaths of pure, clear Tibetan air, that I finally saw my Tibetan Snowcock, a bird I had almost died trying to see. I had at last acclimatised to these dizzying heights. It was a good moment.

Several days later, Lou confided to me that he really feared that I would have slipped away during that night in Madoi. He told me that only one year earlier, he was on a trip with several very fit and experienced mountain walkers from Germany. One man had died from altitude sickness during that trip and apparently his symptoms were not nearly as bad as mine. He smiled when he also informed me that he had begun to make preparations to have my body taken back to Xining. These were the inherent risks of travelling in Tibet and I accepted that this was the case.

I turned forty years of age out on the Tibetan Plateau, and on one high pass I cast blessings to the winds on small pieces of coloured paper, calling 'Axu!' (pronounced 'Ashoo') to invoke the spirits of the mountains to grant my requests. Strangely enough, those requests have been granted. The mountains have been kind to me.

Our journey over the following weeks took us as far as the road went. We crossed high passes without any further issues. I even birded at over 6,200 m (20,000 ft) without a problem.

Then the road just came to an end. We could go no further. We retraced our steps. It was such a great feeling to be given a second chance to see birds like the Black-necked Cranes again. It was such a pleasure to be given a second chance to appreciate the spectacular mountain passes and rivers.

I stood at Erla Pass on the return journey and walked for over two hours across the plains without feeling out of breath. I needed to do that. I needed to let the mountains know that I had survived. I needed to remind myself that it was far from my time to shuffle off this mortal coil.

Just over six months after I returned from Tibet, I left my good, secure, pensionable career in Eircom to set out on a new path in life.

I had a lot of living to do yet.

One year later, I was very saddened to hear the news that Lou was killed in a car accident while he was guiding birders into the mountains of Tibet. He was a superb guide, great company and a real gentleman.

Chapter 25

CAD ATÁ TÚ A DHÉANAMH?

Since walking away from Eircom, I had made my birding my new career. I was now working with schools, sharing my knowledge and passion for birds with Ireland's young birdwatchers of the future and running workshops and courses for adults. I was also now becoming a regular on the *Mooney Goes Wild* nature show on RTÉ Radio 1. I was also in demand to guide birding tours and visitors to Ireland. Birds and birding was now my life.

And so it was that, in the summer of 2005, I was guiding two visiting US birders, Sue and John, on a five-day birding trip. They were a lovely couple, easy company, good birders and good conversationalists. They had retired and were now in their early seventies. Both had Irish ancestry and this was their first time to visit Ireland. It was a trip they had planned for years.

We had birded around Dublin, Wicklow and Wexford and it had been an excellent few days. On the afternoon of our fourth day, we pulled into Banagher, Co. Offaly. Having checked into our B&B, and after grabbing a quick bite of food, we set off down a long grassy path that ran parallel to the Shannon River. It was lovely balmy night. Hundreds of acres of rich wild hay meadows stretched out in front of us. A 'drumming' Snipe was flying around doing his acrobatic aerial displays, falling through the air, allowing his rigid outer tail feathers to make a whirring sound. A Cuckoo called off in the distance.

We stood in the glow of twilight. Then I heard a distant sound. It was a long way off.

'Shhh … there's one calling.'

We strained our ears to listen but we couldn't hear a thing.

Then an explosion of sound came from the field right beside us.

'Kerrrx-kerrrx … kerrrx-kerrrx … kerrrx-kerrrx.'

The sound echoed around us. In fact, it was so close that the bird's call was almost vibrating through my body. I sighed with relief (only bird guides will know that sigh) and smiled at them.

'Corncrake!'

The bird called again and then another called from behind us. The birds started calling to each other. It was magical.

Suddenly Sue fell to her knees, sobbing uncontrollably. I didn't know what was wrong. I looked with real worry at John, but he was smiling. He waved his hand to me to say that all was okay.

I stood a little away. Sue sat down with big tears rolling down her cheeks. She was crying and smiling at the same time. Eventually, she stood up, wiped her eyes and laughed.

She came to me and gave me a big hug.

'Thank you, Eric,' she said. 'Thank you so much.'

As we celebrated our Corncrake over a refreshing pint of the black stuff later that night, Sue explained that her grandmother had left Ireland as a young child. Almost seventy years later, a five-year-old Sue climbed onto her 'Granma's' lap and asked her what she remembered about Ireland. Her grandmother then told her about the Corncrake.

Her grandmother's abiding memory of Ireland, one that she carried with her throughout her whole life, was the call of the Corncrake. She had never returned to Ireland. That day, five-year-old Sue promised her grandmother that she would go to Ireland some day and hear a Corncrake for both of them. Sue had made good on her promise.

Corncrakes are long-distance migrants, arriving in Ireland in the spring having spent the winter in southern Africa. It really is hard to believe that these chicken-like birds, which prefer to run rather than fly, actually migrate across the Sahara Desert, over North Africa and into Europe before flying across the open sea to reach our shores. Ireland is a long way from the grasslands and savannahs of southern Africa.

When they reach Ireland, they spend the summer in the rich hay meadows, where there is cover for them to nest on the ground and a rich biodiversity of insect life for them to feed themselves and their chicks.

The birds call mostly at night and many older country people recount that they were a real nuisance, and that it was hard to get a good night's sleep with 'the birds shouting from every field' (as one man described it).

Corncrake males are very territorial and their rasping calls are warnings to other males to 'keep out'. They come into Ireland just before the females and they do their best to lay claim to prime real estate before the females begin to arrive. The calls also attract the females.

Throughout all of Ireland, the rasping call of the Corncrake was *the* sound of summer in the countryside and it was this sound that Sue's grandmother had carried with her for her life. It seems such a shame to think that the current generation who are leaving Ireland to start a new life abroad will not carry the memory of the call of a Corncrake with them. It seems that the magic of Corncrakes will be lost to future generations of Irish, at home and abroad.

When I was born, in 1961, Corncrakes were still to be found in every county in Ireland. It's hard to believe that now. I even remember my sister, Clare, telling me that her teacher had got her class to stay quiet so they could hear a Corncrake calling in the grass in the school grounds. Corncrakes were breeding in Finglas in the 1960s.

However, even in the 1960s the species had been in very sharp decline for many years. Changes to farming methods had taken a huge toll on these birds. Nesting Corncrakes are reluctant to fly and will not abandon their nest or chicks. Their main method of escape is to run. In the 'good old days', when farmers cut hay meadows with scythes, the cutting process was slow and did not pose any danger to the nesting birds. If a farmer came across a Corncrake on her nest, he would cut around her and not disturb her. Horse-drawn mowers also did not pose a threat. If the Corncrake had chicks, the slow pace of the mower meant there was enough time for the whole family to escape before disaster struck.

However, as cutting processes modernised and tractor-drawn mowers were used to cut hay meadows, the nesting birds rarely escaped. Of course, for convenience, fields are cut from the outside in and often the female with her chicks might be trapped in the final swathe of grass in the centre. The whole family would be killed in that final cut.

As we slowly moved to more intensive farming, pesticides became the most common method of controlling insects and so the decline of the Corncrake quickened with this assault on their food. Almost the final nail in the coffin was the development of silage cutting, where grass is cut many times during the summer season. At least when we allowed hay to grow, the birds had some chance to raise one brood. Hay was not cut until late in the summer so this often gave a pair of Corncrakes enough time to nest successfully.

Additional pressure comes with hunting along the North African coast and the ever-expanding sands of the Sahara. Even on their migrations, the birds face great dangers. Finally, their southern African wintering grounds are also under great pressure from continuing human encroachment and changes in land use.

So, even by the late 1970s, people realised that Corncrakes faced a very bleak future. They were considered to be on the slippery slope to extinction in Ireland. By then, they were confined to the more remote western regions, like the Mullet Peninsula in Co.

Mayo, where farming methods were still more traditional. These were the Corncrake's last strongholds.

I actually had my first Corncrake experience on the Mullet Peninsula. It was June 1982 and I was with Ronan. It was one of those special birding moments in my life that are etched into my brain. However, another Corncrake moment on the Mullet will also remain with me for the rest of my life, but for quite a different reason.

It was June 1987, and I was birding with Mick. It was a warm sunny day. We had spent over six hours and driven the length and breadth of the Mullet trying to find a Corncrake. Earlier, we were even fooled by a Starling that did the most perfect imitation of a Corncrake call. It was better than a real Corncrake. But now, we had pinned one down in a beautiful large hay meadow close to a national school on the southern part of the peninsula. Insects hummed and there were Grasshoppers everywhere. In there somewhere was a Corncrake and he was responding to the sound of my good old trusty Corncrake bones.

I had spent a long time during the winter crafting these bones. They were small pork rib bones that I'd scrubbed clean and dried for a couple of months. Then, with a small hacksaw, I had carefully cut out neat serrations along most of the length of one. With a Stanley knife, I narrowed the edge of the other bone. At each point, I checked the sound they made as I ran them across each other. Like a master craftsman, I made minute adjustments until I felt that at last I had the sound and the pitch just right.

Today, I was putting them to the ultimate test and they were proving themselves to be a 'good set of bones'. Once again, I rubbed them against each other several times and waited.

'Kerrrx-kerrrx … kerrrx-kerrrx … kerrrx-kerrrx,' came the reply.

Of course, it was the very noise of my 'Corncrake bones' that had fooled this Corncrake into thinking that a rival male was challenging him. It is also worth stating that such 'luring' is now illegal and luring in any way should never be used. But in the 1980s everyone had a set of Corncrake bones; it was the 'done' thing.

I was lying by the side of a ditch, looking into a swaying sea of grass. Mick was lying just up from me and we had been in this position for over two hours. We were like two soldiers preparing to go over the top at the Somme. It was a large field and we had hoped the bird might have flown up to give us an indication of where it might be. Instead, this bird was slowly (and I mean slowly) walking towards us.

We waited and 'boned'. It called back and walked towards us. We waited. It called. We scoured the tops of the grass in the hope of catching a glimpse of a head looking around. We did not see it. We listened for the sound of the grass rustling. We heard nothing. We remained still, not moving a muscle except to use the bones again.

Slowly the bird came closer. I could now almost imagine seeing the grass move as it approached. I looked at my watch. It was 2.25 p.m.

'It will be worth it,' I reminded myself. I had my camera all set up. I had checked my settings. I was primed to capture my first Corncrake photograph.

As the seconds ticked by, the bird came closer. It was calling from only a metre away. It felt like torture watching every blade of grass for movement. The bird was now calling constantly. I could feel the sound in my own bones. It was coming closer and closer, calling as it approached. I could almost touch it. Was that it? That movement? I picked up my camera and focused on where the grass had moved. I could see the dark shape of a bird carefully walking through the dense vegetation. I focused my camera again. Was that a black beady eye staring at me through the grass? The shot was almost within my grasp. The bird's head was about to appear above the grass. It was so close … and getting closer …

Then, suddenly, and out of nowhere, we found ourselves surrounded by a group of chattering kids. School was over and they had seen us lying across this ditch. Their curiosity had got the better of them and they'd come running up to us – just as the bird was about to show itself and just at the very moment I was about to capture my first Corncrake photograph.

'Cad atá tú a dhéanamh?' one little lad asked me in Irish, meaning 'What are you doing?' It was a Gaeltacht area.

We tried our best to ignore them. We watched for movement in the grass below. The Corncrake was now silent. The Corncrake was now very still.

I boned again. There was no response.

'Gabh mo leithscéal [excuse me] ... cad atá tú a dhéanamh?' this little lad asked again very politely, but much louder in case I hadn't heard him the first time.

I saw the grass move as the Corncrake fled back into deep cover.

I turned around, feeling the tension of hours of waiting for this wonderful moment etched on my face.

I saw about ten kids looking at me. The little fella who had asked the question was about ten years old, with brown curly hair and glasses. I looked around at the group of curious kids.

'Would ya all ever go and fuck off!' I said in a broad Dublin accent.

My request was delivered with venom. I couldn't believe what had come out of my mouth.

The kids, like the Corncrake, fled into deep cover. Neither was seen again.

Mick and I looked at each other. We sat silently for a moment before we collapsed into laughter. The poor kids. Instead of us explaining to them how important Corncrakes were, I had 'effed' them out of it.

As I've said, I now work with national schools across Ireland and enjoy encouraging kids to take an interest in the beauty and wonder of our birdlife. I am still horrified by my reaction that day, but ... there are no 'buts'. If any of those kids are now reading this, I unreservedly apologise.

For the record, I eventually managed to capture some nice images of a Corncrake in 2009, twenty-two years after that encounter with the kids on the Mullet Peninsula. It was a long wait but worth it in the end.

Chapter 26 ✦

A MOMENT OF RECKONING

There are some moments in life that make you stop and take a long hard look at yourself. There are days that totally change things. In my birding life there was one day that changed everything for ever. This was Saturday, 14 October 2006 and I was on Cape Clear Island.

I had spent the last few days of September and the first few days of October on Cape Clear. It was a great birding week that included an Eastern Olivaceous Warbler (a seriously good tick) as well as the likes of Rose-coloured Starling, Wryneck and Richard's Pipit. For a short stay, it was a good haul of 'eastern' rare birds.

On 12 October, I was back down in Co. Cork. I was giving a talk that evening to launch a new nature trust group near Douglas and would be hosting a migration workshop for beginner birders that weekend in Rosscarbery in West Cork. The migration weekend included a trip out to Cape Clear on the Saturday. I was looking forward to the few days ahead. It was great to be able to combine speaking, guiding and birding during peak migration times in one of the best birding counties.

As I arrived in Cork, Owen Foley, a young Dublin birder who had learned his trade with the Tolka Branch of Birdwatch Ireland, contacted me to let me know that there was a juvenile Red-backed Shrike and a juvenile Woodchat Shrike together in the same field on the Old Head of Kinsale. These are rare European birds to Ireland and to have them side by side for comparison was a real treat. It was a perfect start to my few days away, so I headed straight out to the Old Head. It's always great to arrive at a place and to see a birder (or birders) there ahead of you. It means you don't have to start from scratch and search for the bird. Owen had the shrikes lined up.

I walked over to him and he greeted me by following a golden rule in birding: he stood aside from his scope so that I could see the birds through it. This might sound like a simple thing, but it's such an important part of birding etiquette. It is something that I do automatically. If I am watching a rare bird and a birder arrives to see it, I stand back and let him or her look through my scope. For a twitcher who might have travelled for hours, the relief of just getting your eye on the bird is indescribable. Once you've actually seen it, then you can relax, set your own scope up and enjoy the bird at ease. It is something I do and something all birders do. It is based on the 'do unto others as you would have done unto yourself' ethos. We have all given this gift to other birders and we have all benefited from this gift over the years. It doesn't matter whether you know the birder or not; it's just the 'done' thing. If you know the birder, then it's even better: you have shared the moment with a friend. It's a warm feeling to share a good bird with others.

It was a lovely afternoon and, after having enjoyed the shrikes and thanked Owen for staying with the birds until I had seen them, I birded around the Old Head for the rest of the day. Then I headed into Douglas where I was due to speak on birds that evening. After a quick bite to eat, I arrived at the venue and was setting up when I got a call. It was from Cape Clear. A Baltimore Oriole had been found there late that afternoon.

This was Ireland's second record for this colourful North American, Blackbird-like species. Amazingly, Ireland's first

Baltimore Oriole had been found only five years earlier at Baltimore in Co. Cork. That was some coincidence: a Baltimore Oriole in Baltimore. I had twitched that bird immediately, driving from Dublin to Cork as soon as the news broke. I dipped on it, but so did everyone else that day. It was seen briefly the following morning. Now, here was another Baltimore Oriole on Cape Clear and I happened to be in Cork.

I updated the BINS line with a red alert and sent out over 100 red-alert texts to the birding community. I then hurriedly completed my set-up and gave my talk to a very warm and appreciative audience.

By the time I had completed my talk, a boat had been chartered to go to Cape Clear early the following morning (Friday morning). While this was tempting, I really could not risk going to Cape in case I could not get off the island again for some reason. I was commencing my migration workshop weekend on Friday evening in Rosscarbery with a 'meet and greet' and a talk. Besides that, I knew that if the bird was going to stick around, I would have a chance to see it on Saturday. It was a brave decision but I had a commitment to the thirty people who had booked in for the workshop.

The following day, I birded my way westwards. News from Cape Clear was that the oriole had not been seen yet that day. I found a delightful Yellow-browed Warbler (from Siberia) at Galley Head. As I approached Rosscarbery, I pulled in at a quiet spot on the western edge of the estuary where I had a good mobile-phone signal. I was due to do a live radio interview for the *Mooney Goes Wild* show on RTÉ Radio 1. As I sat waiting to give my spiel on radio about bird migration, I glanced out the window to look at a group of Redshanks feeding below. There seemed to be a slightly smaller, more elegant 'shank' with them. As I came live on air, I was raising my bins to see a dainty Lesser Yellowlegs (the North American version of our Redshank) just below me. I actually found it live on air.

It was great fun, and by the time I had finished the interview over forty listeners who had been driving home to, or through,

Rosscarbery that Friday evening swung by to see the bird and to meet me. It was an impromptu twitch by non-birdwatchers. It was great to be able to show such a rare and special bird to total beginners.

By the time the workshop's evening 'meet and greet' talk commenced, the positive news from Cape Clear was that the Baltimore Oriole had been re-found in a garden along the Low Road. It looked settled there. Later that evening, all of my workshop guests were standing enjoying excellent views of the Lesser Yellowlegs.

Saturday, 14 October dawned bright and sunny. Before breakfast, I got a text from Steve Wing, the warden at the Bird Observatory on Cape Clear. (Incidentally, isn't that a great name for a warden of a bird observatory? And a previous Cape Clear Bird Obs warden was called Dave Bird. I kid you not.)

Steve's text was short and sweet, and contained exactly the news I hoped to get:

'Morning, Eric. Great news. The oriole is still here. See you soon. Steve.'

Following breakfast, we had time to take another look at the Lesser Yellowlegs before the bus arrived to transport us to Baltimore for the 11 a.m. sailing to Cape Clear. We arrived and boarded the boat. There were already a lot of anxious twitchers on board and there was the usual air of tension that exists before twitchers see their bird. As a twitcher, I understood that very well. I had that feeling myself, but my focus was on giving my workshop guests a great birding experience (and hopefully getting a tick while I was at it). Most of the birders on board had received my red-alert text and I spoke with them all briefly. I kept my group away from them. Few hardcore twitchers want to be distracted by non-hardcore twitching talk.

In the North Harbour of Cape Clear, Steve was there to meet my group. The birders dashed off up Cotters Hill. If I were not guiding my group, I would have been with them. Steve and his partner, Mary, were wonderful hosts and we all enjoyed a tour of the Bird Obs before we commenced a walk around the island.

We stopped off at Cotters Garden and wallowed in the glory of a little Yellow-browed Warbler that showed itself well to everyone. I had managed to show my migration group birds from opposite sides of the world in one day. This little warbler was from Siberia, while the Lesser Yellowlegs was from North America. There were also lots of Chiffchaffs around the garden, which offered great opportunities for comparison with their rarer Siberian cousin.

Then the news came that I hoped to hear. One of the birders came into Cotters and let us know that the Baltimore Oriole had been seen several times and was showing quite well on occasion. With the main rush over, and all of the birders now calm after having seen the bird, Steve and I brought our group along the Low Road to the garden where the oriole was present. There were about twenty birders still there, standing in the only area from where the garden could be viewed.

As my group arrived, there seemed to be tension among the birding tribe. It was very obvious from the lack of acknowledgment of my group (and of me) that the hardcore birders did not want a gang of beginners hanging around them. Not one person smiled or engaged any of my group in conversation. Not one of them moved to allow us our chance to look for this mega bird, even though every one of them to a man had already seen the bird. Just about every birder standing there had received a red-alert text from me.

I chose to quickly bring our group on. I wanted to shelter them from such a negative and intense atmosphere. I wanted them to enjoy the birding experience of Cape Clear. Thankfully, for all of these people, seeing the oriole was not important. They were enjoying a wonderful day on Cape Clear and we birded the west side of the island. I even took time to drop in and introduce them all to Mary Mac, who gave us her usual warm welcome. I had retrieved the situation.

As we returned down the High Road, I could see that the tribe of birders were still hanging out around the garden on the Low Road. They were chatting and smiling. The bird was elusive, but they had all seen it well several times. Over lunch, my group voted that I should have some time off. They knew how important seeing the oriole was for me. They wanted me to see it. Steve assured me that he would look after them.

I dashed back up to the Low Road and joined the gang of about twenty birders who were still there. As before, they occupied the whole viewing area, so I got into the best available position from where I could see most of the garden. Everyone was watching and talking. I was the only person who had not seen the bird. Every birder there knew that. Then the Baltimore Oriole appeared.

'There it is again!'

'Where?' I asked.

I couldn't see it.

'Fuck! That's a great view ...'

'Where is it? Can someone tell me where it is?'

I still could not see it.

'Wow! That's superb.'

I lowered my bins and saw that all the birders were looking through their scopes at a point in the garden to which, from my position, I had no line of vision. The bird was apparently sitting out in the open, but it was impossible for me to see it from where I stood.

'Lads, can I get onto someone's scope please? I can't get it from here,' I said loudly.

They all heard me but no one looked up. No one moved.

'Lads, I can't see it. Can someone please let me grab a look in their scope?'

My pleas fell on deaf ears.

Some of these guys were birders I had known for years. Some were birders that I had invited to look through my scope to see a mega rare bird in the past. I could not believe that this was happening. What had happened to Irish birding? Had twitching become this nasty dog-eat-dog activity? Where was the camaraderie among fellow birders?

Again I pleaded.

'Lads, I can't see the bird. I need the bird. Can someone please let me look in their scope?'

With these words, the bird moved and the moment was lost.

I stood in silence. The episode had lasted about thirty seconds. Count up thirty seconds. It is a long time for a bird to show itself. It is more than enough time to allow a birder, standing shoulder to shoulder alongside you, to stand back and let you take a quick look through their scope to see a bird. Thirty seconds is a long time in a birder's life. In my life, it was a defining thirty seconds.

The bird went to ground so I walked down into the garden, where I met Jim Dowdall and Paul Kelly. They smiled at me.

'You must have got great views of it that time from up there,' Paul said.

I explained to them what had happened. They were horrified. Together we spent another short while looking for the bird, but we couldn't find it. They both worked very hard to help me find it, but to no avail. I appreciated their help more than they realised.

I returned to my workshop group. They were disappointed that I had not seen my bird. We had a great few hours. As part of Steve's work as warden of the Bird Obs, he frequently caught and ringed migrant birds (placing small, lightweight, coded rings on their legs which allows us to monitor their movements). That afternoon we actually caught and ringed that Yellow-browed Warbler. It weighed just 4.2 g. Imagine a bird that tiny and light had flown from Siberia to Ireland.

We caught the boat back to Baltimore that evening. I left the island without having seen the Baltimore Oriole. On the journey back across Roaringwater Bay, I sat quietly, reflecting on what had happened that afternoon. I knew that hardcore birders were averse to meeting beginners at times. I never understood that. It was as if these birders had forgotten that they too were beginners at one time. I had never forgotten those moments in my early birding years when someone went out of his or her way to show me a bird. Hardcore twitchers can have a very short memory.

Why did no one move to allow me to see that Baltimore Oriole? Was it that they resented the fact that I was making money from being a guide – that old BINS problem yet again? Surely not? While there was still some lingering resentment among a very few, those heady, heavy days of the early 1990s were long gone. However, preventing me from getting a tick had brought this to a new low. Was it that they enjoyed having one bird over me? Did they believe that because I was guiding a group of beginners, I was now no longer part of 'the gang'? Did I really want to be a part of any gang that behaved like that? Yet, many of those same birders had first learned of the bird by a personal red-alert text from me. Some of these were fellas I spoke to on a weekly basis. Some of them were people who rang me up if they needed directions or updates. These were my birding colleagues. These were people I would often refer to as friends.

I felt numb.

That evening, we had a lovely group meal in the hotel at Rosscarbery. I enjoyed it. My group were so warm, and it felt good to be in their company.

The following morning we completed our workshop. The feedback was great. Everyone had seen some great birds and had enjoyed the Cape Clear experience. It had been a great success. By lunchtime, I was back by myself at the estuary at Rosscarbery. My phone buzzed. It was a text from Robert Vaughan, another young Tolka Branch birder. I read the text.

'Baltimore Oriole still present in garden along Low Road, Cape. RV.'

Moments later, my phone rang. It was Steve.

'The oriole is showing well again, Eric,' he informed me. 'Why don't you come back and we'll spend the day trying to see it?'

Steve was very upset that I had missed the bird. He was so encouraging. However, today, a new 'me' had emerged. Seeing the Baltimore Oriole had no meaning for me now. It was no longer important. I almost did not want to see it. Those thirty seconds had utterly changed my birding life, and my life.

Twitching was great. I had had so many adventures. I had seen so many birds. I had shared the company of so many wonderful people. Now I was choosing to step away. To take a big step away. I was ready for a new journey where the pull of the tick was no longer the main draw of birding. I had already started on this journey, but today was *the* day when I was finally released from my addiction to twitching. I felt as if a great weight was being lifted off my shoulders.

'No thanks, Steve, my old friend,' I replied. 'I really appreciate that, but today I'm going birding.'

I walked across the road and enjoyed watching the Lesser Yellowlegs, which was still showing well. Lots of birders, coming off Cape Clear, stopped off to look at the 'legs'. Ironically, one birder, who was part of the oriole group, was in a hurry, and so he was very happy to take a look at the bird through my scope. He got his 'year tick'. It really is a warm feeling to share a good bird with others.

Later that day, I returned to Rosscarbery and found a juvenile American Golden Plover among the large flock of Goldies on the estuary. I was delighted. There were two good birds on the estuary and I had found both. Many birders twitched the American Golden Plover and the Lesser Yellowlegs the following morning en route to Cape Clear to see the Baltimore Oriole, which remained until 19 October.

I often wonder what might have happened had one birder done the right thing that day and let me see the Baltimore Oriole through his scope. Would I be the birder I now am? Would I be where I am now?

I can't really answer those questions.

I still enjoy seeing new birds. I still enjoy getting a tick. I still enjoy twitching, but it's not the same as it was before. It is no longer my driving force.

In those thirty seconds on that afternoon on Cape Clear, my birding life changed entirely. The hardcore twitcher in me was lost in those moments of reckoning and a new me emerged: 'Eric, the Birdman', the birder you see today.

Chapter 27

DEANO

Wednesday, 29 August 2007 was one of those ordinary days in the week. Two days earlier, I had been out at sea looking at seabirds off Co. Kerry with the 'Kerry Lads'. I had travelled down to Kerry on Sunday with Anthony McGeehan, who had parked at my house having driven to Dublin from Belfast. Journeying with Anthony was always both entertaining and educational, with tales of the past mingling with cutting-edge bird identification discussions.

The pelagic trip itself was good, the highlight being a superb Wilson's Petrel that danced off the back of the boat for over thirty minutes. It was the saviour of the day since, until then, we'd only seen small numbers of shearwaters and Storm Petrels along with a single Arctic Skua. However, I didn't mind. I had just bought a new digital SLR camera and was enjoying learning how to use it (up until then I had been using slide film). I was happy with my shots and had even managed a few decent ones of the Wilson's.

On Wednesdays, if I was not away, I always went up home to have dinner with Ma and Da and to have my weekly catch-up with them. We'd sit and chat for a few hours about everything. This

Wednesday was no different and, at around 2 p.m., I got myself ready for my weekly visit. I had printed off some pictures of the Wilson's Petrel to show Da. He was always interested in the trips I went on and the birds I saw. He wanted to know where the birds had come from and how they had got to Ireland.

However, before leaving home there was one major job to do and that was to round up the cats from the garden.

Cats?

Yes, cats. At this stage I had four cats. Each was a rescue cat and each spent their life in the tradition of Quasimodo – in the company of many bells. The bells announced their every movement. This was important, as there were bird feeders and nest boxes in the garden, and nesting birds in the trees. The cats had given up even trying to catch a bird and instead became lazy. They all enjoyed nothing more than to sun themselves and sleep. All, that is, except Deano, who was the adventurous one.

Deano was just about the funniest character of a cat that ever walked the earth. I had spotted him as a tiny ball of fluff walking along the white line in the middle of the main Killarney to Cork road three years earlier. Cars were missing him by inches. I had to do something, even if it was to shoo him off the road. I pulled up and ran back along the road as another car zoomed by, this time the force of the wind knocking the little kitten over. When there was a gap in the traffic, I ran out to this pathetic little creature. He sat down and meowed up at me. What could I do? I picked him up and darted back to the safety of the car. He instantly snuggled into my jacket and purred. I was hooked. On the way home, I stopped and bought a packet of ham and he wolfed it down. From that moment on, he associated the rustle of a ham packet with a treat and would always come running from wherever he was.

Deano was white with black patches on his face and back. His tail was banded with black stripes. When he ran, he always reminded me of a Ring-tailed Lemur. When I found him, the poor little fella was in a bad state. His right eye was infected and his ears were badly sunburnt. Over the following days, all of the fur fell off his ears and never grew back. As a result, Deano always needed to have sunscreen on his ears. His eye never recovered and, when he was old enough, he had that eye removed (a bad day for poor old Deano as he also had two other circular vital parts removed at the same time). Despite all this, Deano was a real character. There was not a single day on which he failed to make me laugh. I think he believed that he was a dog. When people arrived at the house, the other cats would slink away and hide, but Deano would come running up to meet and greet them. Even people who hated cats liked Deano.

So, on this Wednesday afternoon I gathered up Princess, Frigger (don't ask) and Issie within seconds with the promise of a treat. I went outside and called Deano. No sign of him. I called again but this time rustled a packet of ham. This always worked. However, this time, there was still no sign of him. This was not like him. I gave him a few more minutes and called again, but Deano did not appear. I glanced at my watch. I was a little late and so I decided to leave him outside. It was 2.15 p.m.

I went up home and let myself in. Ma was in the kitchen and Da was upstairs. Ma was putting dinner out but she had a strange look on her face. It was look of worry.

'Howya, Ma. Are ya all right?' I asked.

'I'm worried about Da,' she replied. 'He was at the doctor this morning.'

She told me that during the morning she had noticed that Da was looking a little yellow. The whites of his eyes were yellowish and his back and stomach were the same. She was so worried about this that she had taken him over to the doctor that morning for a blood test. The doctor didn't like the look of him and had sent the blood off immediately. She was waiting for the doctor to call with the results.

Moments later, Da arrived in the kitchen with a big smile on his face.

'They call me mellow yellow,' he sang. There was nothing wrong with his sense of humour.

'So, what do you think?' he asked. 'Ma thinks I look yellow …'

There was no doubting that he did look yellow. We lifted his shirt and looked at his back and stomach. He was indeed looking a bit jaundiced.

'So?' he said. 'How yellow do I look?'

'Eh, you know what an embalmed corpse looks like?' I replied. 'Well, you look like an embalmed corpse!'

Da burst out laughing.

'Well, thanks very much. You wouldn't want to be looking for compliments around here!'

We sat and enjoyed a lovely dinner and we chatted about him being yellow, and how the webs of a Wilson's Petrel feet were as yellow as he was. Da started washing up (he always insisted on washing up) and, when he was up to his elbows in suds and Ma was making tea, the phone rang. I answered it. It was the doctor. He sounded very serious. The initial results showed that Da's liver was not functioning properly. Da needed to get into hospital straight away. The only way was to bring him into the Accident and Emergency Department (A&E) in the Mater Hospital. In the meantime, the doctor would inform A&E of Da's impending arrival. I put the phone down and returned to the kitchen.

We poured the tea. As we sat enjoying the brew, I informed Da that, when he had finished his tea, we were off to the Mater. He thought I was joking at first. It seemed so ridiculous that a few minutes before he was doing the wash-up and now he had to get to hospital immediately. The poor man really did not see that coming. None of us did. Ma packed a small overnight bag and I called a taxi. It would be easier to get a taxi and leave my car parked at home. So, off we went to the chaos of A&E.

I will not go into too much detail about the dreadful hours spent sitting with my 83-year-old father on hard chairs in a waiting room alongside a few drunks who were asleep on the floor. Da was tired and a little weak. Hour after hour passed. Even other patients, when they were called, tried to convince the doctors that they should see this elderly man first. Their requests fell on deaf ears.

Da was eventually seen but only after I lost my temper a little. I asked a doctor where the humanity and caring of the hospital had gone. Where had respect for the elderly gone? I reminded him that my father had worked for almost forty years in this very hospital. He deserved to be treated with a little dignity. Everyone deserves to be treated with a modicum of dignity.

In the meantime, Paul had been in touch and was coming down to join us on our vigil. It was now after 8 p.m. and I was well aware that I had not returned home since I'd left for Ma and Da's earlier. I had visions of poor old Deano sitting waiting at the back door, so I asked Paul if he could drop by and let him in before coming to the hospital. However, when Paul arrived at A&E, he informed me that there was still no sign of Deano. That really was unusual. Paul also told me that he had met my neighbour Fran and had told her that I had been in A&E all day with Da. He asked if she would keep an eye out for Deano. Fran was not a cat lover but she liked Deano. She promised she would.

At long last, Da was brought into the treatment room and seen by several doctors. A number of initial examinations were carried out and then, at last, Da was given a trolley to lie on. What a sad reflection of our health service when even getting an 83-year-old man onto a trolley can bring such great relief for worried family members. I went back into the cubicle where Da was now resting on his trolley. We were told that nothing more would happen again until the morning so, as Da began to fall asleep, we left A&E. I kissed him goodnight.

Paul left me to Ma and Da's to collect my car. We sat for a few minutes chatting with Ma before she ushered us out the door, telling us to go home and rest.

I arrived home expecting to see Deano at the back door, screaming his head off. I was feeling weary and I really was looking forward to seeing his cheery, cheeky face. But he was not there. I went into the back garden and called him. I rustled ham packets. I walked my housing estate in case he had gone wandering. No sign.

I eventually got to bed well into the early hours of the morning. Sleep did not descend. Instead, I lay awake thinking of Da lying on his trolley in the Mater. I hoped that he was okay. And I wondered where the hell Deano might be. I got up and called him several times during the night. Each time I went downstairs, I really thought he would be sitting at the door waiting to be let in. He had never gone missing before and I was now very worried.

So, in an attempt to occupy my mind and do something constructive, I designed a 'Have you seen Deano?' poster. I printed off about 100 of them and, at 7 a.m., I started to walk around my housing estate and around the Willows (the estate behind where I lived), pushing these posters through people's letterboxes. I met several people getting into their cars heading off to work. I think most thought I was a nutter, but several animal lovers wished me luck. As I walked away from one house in the Willows, a woman suddenly ran out the door and came after me with the poster in her hand.

'It that your cat?' she asked, pointing at the poster.

'Yes. Have you seen him?'

The look on her face told me that this was not good. She explained to me that she too had a black-and-white cat. A neighbour had called to her home yesterday to ask if she was missing her cat because a black-and-white cat had just been killed by a car. My heart sank. She saw that in my face and brought me to her neighbour's house. They had picked up this cat and knew that he was obviously someone's pet. They had him in a black bag and were waiting for him to be taken away by the City Council. The bag was in the garden. I went out and stood over the bag for a minute before opening it and looking in. There, curled up as if asleep, was a cold but perfect Deano. He had been killed just after 2 p.m. the previous day. Whilst I had been calling him, he was lying dead on the road. All manner of thoughts ran through my mind. Why didn't I call him earlier? Why did I let him out?

I thanked them so much for their respect for Deano and assured them that he deserved that respect. I carried him home. I felt sick and I couldn't hold back the tears. My little friend was dead. I reached home and sat down in the back garden with Deano on my lap. I was alone and I sobbed like a three-year-old.

With that, the doorbell rang. I wiped my eyes, gathered myself together and answered the door. My neighbour Fran, to whom Paul had spoken the night before, had seen my car parked. She was calling in to see if there was any news or updates. One look at my tear-stained face and red eyes told her all she needed to know.

'Oh, Eric. Are you all right? Is it bad news?' she asked.

'He's dead,' I answered, beginning to sob again.

'Oh, my God … I'm so sorry, Eric.'

She gave me a huge, warm, comforting hug.

'What happened?'

'He was hit by a car.'

'Oh, Jesus Christ … he was killed by a car. Where did that happen?'

She was clearly in shock.

'In the Willows,' I answered. 'But at least I found him. I have him in a plastic bag in the garage.'

'You have him in a plastic bag?'

The pitch of her voice was rising.

'Who is in a plastic bag? What are you talking about?'

'Deano,' I replied. 'I am talking about poor Deano.'

'Ah, feck Deano!' she exclaimed. 'I was asking about your Da!'

'Oh, Da … eh, Da … We're not sure about Da,' I said. 'We'll know more later on. Thanks for asking.'

She gave me another hug and sympathised with me about the loss of Deano.

I shut the door and found myself laughing and crying all at the same time. I laughed at the madness of such a mix-up. I cried for my pet. I cried for my Da, who was lying on a feckin' trolley in the Mater. I cried for the feeling of utter loneliness that seemed to envelop me at that very moment.

Paul came to my home soon after and together we buried Deano in the garden. As an animal lover, he understood my heartbreak.

I thanked my little friend for making me laugh so much in the three short years I had known him. Even when he was dead, the total hilarity of the mix-up with Fran, my kind and caring neighbour, was his final gift to me.

With that done, Paul and I went back into A&E to face what the world had to throw at us.

Chapter 28

THE BEST AUTUMN WE'VE EVER HAD

Autumn 2007 was the best autumn I have ever experienced in my life. During the course of September and October, I managed to get three new Irish ticks. This, for me, was undoubtedly the best birding autumn for years.

However, it is not for the ticks that I rate autumn 2007 as the best of my life but for totally different reasons.

It started with a phone call I received towards the end of the first week of September.

It was Ma.

'Hi, Eric. Are you busy?' she said.

'No, fire away,' I replied.

'I have Da here and he wants to talk with you.'

With that, Da came on the phone. It was a call I was not prepared for. Da was still in the Mater Hospital undergoing a series of tests. It was strange that he wanted to speak to me, especially since I was due to call in to see him in a few hours. I sensed this would be one of those phone calls.

'Howya, Da. How's things?' I said.

'Well, now,' he began, 'that's exactly what I want to talk to you about.'

'Okay, Da … I'm listening.'

'You know I've had all of these tests and the like … Well, the doctor has just been here …'

He paused. It was a pause that spoke volumes.

'Well, the news is not good I'm afraid,' he announced matter-of-factly. 'They've told me I have cancer.'

'Is it bad?' I asked. It was a stupid question, but it was the only thing I could think of to say.

'Yes, very bad,' he continued. 'They say I have only a couple of months left.'

His words hit me hard. I had almost prepared myself for the test results not being good, but when my own father said the words … Well, they were difficult words to hear. I found myself swallowing hard. Da sensed my reaction.

'Now listen to me,' he said. 'I'm over eighty-three years old. I've had a good innings. At my age, something is "gonna get ya".' He said the last bit with a laughing tone in his voice.

'I need to say something to you,' he continued. 'I do not want any weeping or gnashing of teeth.' This was a phrase often used in our house and, for fun, we always pronounced the 'g', as in 'g-nashing'. Da included the 'g'.

'So … no maudlin or crying, do you hear me?' he said.

I was so impressed by his words. This was the old Tom Dempsey, logical and matter-of-fact. This was Da looking after me, making sure that I would be okay with the news he had received.

I gathered my thoughts and reminded him of the deal that we had struck many years ago. In fact, it was a deal I had insisted on with both Ma and Da.

The deal was simple: no dying during the autumn bird migration months.

'But, Da,' I found myself saying, 'you can't die yet. It's peak autumn migration time. We had a deal – remember? You're not allowed to die in autumn!'

He laughed. It was good to hear him laugh.

'You're right,' he said. 'I'll try to hang on until November. Would that be okay?'

'Hang on for as long as you can, Da,' I answered.

'Right … I'm going to hand you back to Ma. You know everything.'

Ma came back on the phone and told me she'd talk to me later. There were other calls to make. I hung up. The words and the gravity of the situation came in upon me.

'No maudlin,' I said to myself. 'No weeping or g-nashing of teeth.'

From that moment on, I decided that I would spend as much time with Da as I possibly could. I vowed that I would enjoy his final autumn months.

With nothing more that could be done, Da was soon discharged from hospital and came home. We all decided that he would not return to hospital. He was happier at home. Both Paul and I lived very close by and, between us, we spent a couple of hours each day with him. However, I did have guiding to do and so I continued to head off to various parts of the country in pursuit of birds.

In early September, I was guiding an Irish group in Co. Wexford and had the joy of seeing a Black Kite. This is a very rare species for Ireland and was a much-appreciated Irish tick for me. The same weekend I managed to show my group an Osprey, as well as North American visitors such as Buff-breasted Sandpipers and American Wigeon. Mid-September had me back in Co. Wexford again, this time with some US birders, where we saw lots of birds including North American waders like American Golden Plover (two different individuals), a Baird's Sandpiper and four Buff-breasted Sandpipers. My visitors were more interested in comparing these birds with European waders such as European Golden Plover, Curlew Sandpipers, Little Stints and Ruff. As a guide, I could not ask for more.

In mid-October, I was again booked to guide, this time with two UK birders. I was guiding them around the headlands and wetlands of Co. Cork. I took the long route down to Cork, stopping off in Co. Clare to see my first Irish Buff-bellied Pipit, which gave incredible views as it fed on insects among the piles of washed-up seaweed at Clahane Beach.

In Cork, I saw Ireland's second Blyth's Reed Warbler ever; the first was seen on Cape Clear in 2006. It was my third tick in four weeks. As well as that, we saw another rare European bird in the shape of a Greenish Warbler as well as several Yellow-browed Warblers and a Lesser Yellowlegs. Once more, this was a highly successful trip and my UK birders were delighted with the birds we had seen.

However, my heart wasn't in it. It was elsewhere, as was my mind. I can scarcely remember even seeing some of those birds; if it were not for my notebook, I doubt I could recall what birds (other than the new species for me) I'd seen on those three trips. By this stage I was no longer twitching as much as I used to, but even I would have travelled to Inishbofin, Co. Galway, to see Ireland's first Mourning Dove that had been found by Anthony. Seeing a mega found by a friend is always special.

However, while everyone else was travelling to Galway, I was travelling home to be with Da.

During September and into October, he was strong enough to go out. We'd drive off to somewhere nice, usually to the coast, and just sit chatting about everything. I had decided that I would not just spend time with this man, but that I'd really talk to him. I wanted to tell him how I felt about him. I swore to myself that I would leave nothing unsaid. It was a wonderful opportunity gifted to me. It was a wonderful six or seven weeks.

As October ended and November arrived, Da grew more tired. He was losing weight and getting weaker. So, instead of bringing him out for a drive, I would simply go up home and spend time with him there. He slept a lot and I would just sit with him so that, when he awoke, he knew I was there. I took down the volumes of his carefully scripted poems. He was gifted with a talented hand, and one of his favourite pastimes was to take out his calligraphy set and script his favourite poems onto parchment paper. He then inserted them into clear plastic envelopes before placing them carefully into a folder. I read out many of his best-loved poems and, even though he was getting weaker, he could still recite most of it.

One day, while sitting with him, he turned to me.

'What month is it?' he asked quietly.

'It's November,' I replied. 'Why?'

He smiled with a glint in his blue eyes.

'Well, I've kept my part of the deal,' he said. 'The autumn is over … I can go now …'

I felt big tears rolling down my face. He saw them.

'Remember …' he said, 'no weeping and g-nashing of teeth.'

I smiled.

'So …' he continued, 'was it a good autumn?'

I reached out and held his hand.

'Yes, Da. It was the best autumn we've ever had.'

He smiled and drifted off back to sleep.

As each day passed, he became even weaker. He slept almost constantly. By now he had a morphine pump attached, which dispensed morphine at intervals to ease his pain. The fantastic nurses of St Francis Hospice called on a regular basis to make sure that he was comfortable. They were also great at checking on how we, the family, were doing. Who needs to believe in angels when these nurses are amongst us?

I remember sitting with him on a dull, dreary day in mid-November. He was asleep and I was reading. It was shortly after a nurse had visited. The house was quiet. Ma had gone to the shops. Paul, who had spent hours that morning with Da, had gone back to his house. Golda, the family cat, was tucked up beside him on his bed and was purring quietly. It was just Da and me.

He stirred from his sleep and opened his eyes. He smiled when he saw me. I gave him a drink and asked if he needed anything. He shook his head. He then reached over to me and held my hand.

'We've had some great times together, you and me. Haven't we?' he said.

His voice was weak.

'Yes, Da … We've had some wonderful times!' I answered as I gave his hand a gentle squeeze.

He then looked me straight in the eye.

'Know that I've always been very proud of you,' he said with new strength in his voice. 'You have always done me proud. I am proud of the man you have turned out to be. You've been a great son.'

He paused to catch his breath.

'I have been very lucky to have such a great family,' he continued. 'I am really proud of you all.'

With tears streaming down my face, I leaned over and kissed him.

'And I am so proud to be your son, Da … I have always been proud of you. You know that, don't you?'

He smiled as he nodded.

'Yes, we've had some great times together, you and me …'

He winked at me as he said those words, then closed his eyes and went back to sleep.

I did not know it then, but these were to be the last words we would speak to each other.

On 18 November 2007, at 10.42 p.m., Da left this world. His parting was gentle. It was to his favourite opera music and with his family around him, telling him how much we loved him. I had read poems to him all day, including those of Francis Ledwidge and W. B. Yeats.

Over the course of the day, many people came and went. During the day, there had been several false alarms when his breathing seemed to stop, only for it to start back up again. Everyone eventually departed until it was just us, his family, left.

Then his breathing changed a little.

Ma was in the kitchen tidying up before the night nurse arrived (she was due at 10.45 p.m.). I called her to come in. For some reason, I sensed that this was the moment. It was 10.40 p.m. The clock on the wall marked each second down.

His final breath was one of the most profound moments of my life. It was a comforting sigh, an exhalation. It was the sigh you might give when you're exhausted and your feet are killing you, and then you sit down in a big comfy chair. It was that kind of sigh – followed by total silence.

I leaned over and kissed this great man, told him I loved him and recited a favourite Yeats poem into his ear. I had read that the mind remains active for some time after the last breath and I did not want Da's remarkable mind to leave this world to the sound of tears.

I whispered the words:

When you are old and grey and full of sleep,
And nodding by the fire, take down this book,
And slowly read, and dream of the soft look
Your eyes had once, and of their shadows deep;

How many loved your moments of glad grace,
And loved your beauty with love false or true,
But one man loved the pilgrim Soul in you,
And loved the sorrows of your changing face;

And bending down beside the glowing bars,
Murmur, a little sadly, how Love fled
And paced upon the mountains overhead
And hid his face amid a crowd of stars.

'Good-bye, Da … and thanks for everything.'

The room was perfectly still until suddenly the door of the cabinet that housed the video recorder fell off its hinges and crashed onto the floor. We all jumped. It startled the living daylights out of us.

'Da's energy is still with us,' Ma suggested.

None of us were (or are) religious, but it seemed totally natural that Da's energy might still surround us. I then found myself doing something unexpected, although at the time it seemed the most ordinary, natural thing to do. I got up and opened the window. It was as if I was letting Da's powerful energy back out into nature and back to the universe.

I looked back to the man who was lying in the bed. He was no longer Da. Instead, the body lying there was merely the shell that once housed this incredible man.

Da was gone.

Chapter 29 ✦

A MOMENT BETWEEN FRIENDS

Da's funeral was a wonderful celebration of a great man. It was a humanist ceremony during which we read some of his favourite poems and played his favourite piece of opera, 'Casta Diva', performed by Maria Callas. I read his eulogy. It was my final act for him.

The following months seemed to blur into each other. Grief is a very strange thing. It affects people in so many different ways. Da's death made me feel isolated and lonely. He was one of the few people who 'got me'. He was my confidant, my wise counsellor – a person to whom I could speak on just about anything in the world.

'No maudlin,' I reminded myself.

Despite our family grief, the Dempsey sense of humour still came through. I collected Da's ashes, and I remember the moment I brought them home. Ma and Paul were awaiting my arrival. I carried in the plastic urn that housed his ashes.

'What do they look like?' Ma asked.

'Well, let's put it this way,' I replied, as I opened the lid of the urn to show her the contents, 'you won't be stuck for cat litter for a while!'

We all laughed and cried.

Death is a part of life and life goes on. Over the following year, I threw myself into my birding and photography; it was my best medicine. By now, Mick was living in Co. Kerry and I spent many days visiting 'the Kingdom' and enjoying just being out in the field birding with him, just like the old days. Other good friends rallied around. I spent some fantastic days birding with Anthony in Belfast. In Dublin, I birded hard with Paul Kelly, who always made sure that I had birding company if I needed it. I was also birding with 'the lads': John Fox, Ger Franck and Philip Clancy. These were relatively new birders to the scene and were great company.

I also threw myself into my work. I was running weekly courses in Dublin, working with schools and finishing a new book, *Birdwatching in Ireland with Eric Dempsey*. When that was published, almost a year after Da's death, it was with great pride that I read the dedication:

'For Tom Dempsey, my father and my friend.'

And this was the real truth. Da was so much more than a father; he was one of my best friends too. How lucky am I to be able to say that of my father?

As I have already said, his final breath was one of the most profound moments of my life. The finality of that moment reminded me of my own mortality. It made me take a good hard look at my own life. Again, it's strange to write the word 'life' and not also mean my 'birding life' … the two were, and still are, one and the same.

By now, as I said, I had already taken a big step back from twitching and when, in December 2009, I ceased the BINS line, it seemed like the umbilical cord with that side of my birding had been finally severed. I was leaving the madness behind. I felt I was coming back to my beginnings. While I still enjoyed the buzz of getting a tick, I was no longer addicted to the tick. I felt a deeper sense of wonder surge through my veins again. I was rediscovering the simple magic of birds and it felt good.

Opening up to the magic of birds allows magical moments to happen. One of these moments stands out above them all. It was with a Snow Bunting. I love Snow Buntings. They really are beautiful birds. I never tire of being in their company. My first encounter with Snow Buntings was on 4 November 1979 on the West Pier of Dún Laoghaire, Co. Dublin. I found two birds scuttling like large mice along the broken stones of the seaward side of the pier. Apart from the subtle beauty of their plumage, the thing that always charms me about Snow Buntings is that they are usually engagingly tame.

I have seen many Snow Buntings since those first birds in 1979, but this one special encounter happened at Annagassan, Co. Louth, on 13 February 2010.

It was a cold, crisp, sunny day – perfect for photography. I arrived early in the morning. Annagassan has a small pier, a line of large rocks acting as a sea barrier, a small beach and an area of grassland behind the beach. It is perfect habitat for a Snow Bunting and a female had been reported as being here for over a week.

I put my camera and 500 mm lens (Big Mama, as I call it) onto my heavy tripod and set off along the beach in search of the bird. It took me a while to find her as she was feeding quietly along the area where the grass met the beach. I gave her a wide berth and, settling myself in further up the beach, I waited in the hope that she would come to me. I was sitting down with the tripod just in front of me and my camera trained on the bird as she methodically searched for seeds along the beach.

For me, when photographing birds, I believe that it should always be on the bird's terms. A photograph showing a bird that is looking alert or unsettled because I have crossed into their comfort zone is, in my opinion, a poor image. I prefer to get down low, and, silly as this might sound, allow the bird make the decision as to whether it wants to be photographed or not. If the bird does not feel threatened in any way, it will usually oblige.

And so it was with this Snow Bunting. Slowly but surely, she approached me, getting closer and closer. In fact, she was more often than not too close for photography, unless I was interested in head portraits. This bird was beyond tame. She was something else altogether. She was so charming that I just sat back, forgot about photography and savoured the moments.

For over an hour she fed around me, rarely straying more than 2 or 3 metres from where I sat. It seemed that she trusted I would do her no harm. Here we were – two sentient beings of different species totally comfortable in each other's company on a beach in Co. Louth.

What happened next brought this experience to a whole new level. The beach held lots of shorebirds such as Sanderling, Dunlin, Turnstone and Ringed Plovers, along with a few Pied Wagtails and Meadow Pipits. Suddenly, all of the birds flew up with real urgency. Alarm calls echoed all around. Only a bird of prey would have this effect on birds. I looked around and saw a large female Sparrowhawk moving low and fast towards me along the beach. The Snow Bunting, now about 3 metres from me, was also now on high alert. When danger threatens, Snow Buntings tend to crouch down instead of flying. I expected her to sit low on the ground or scurry into the grass to take cover. However, this bird had another plan.

As the Sparrowhawk came alongside us, the Snow Bunting suddenly ran towards me and came in under the legs of my tripod, almost leaning against my knee. She looked up at me and, if her expression could be translated into English, I am sure it would have been along the lines of …

'Sssh … there's a bloody Sparrowhawk out there. Don't let it know I'm here!'

I held my breath. The Sparrowhawk continued past us and on over the pier and out of sight. It took several minutes before calm was restored to the beach. In those few minutes, we sat together, my Snow Bunting friend and me.

Whilst I was transfixed by the moment, she was clearly so relaxed that she took the opportunity to preen. I could feel her body and wings against my leg as she scratched and fluffed herself up before emerging from under the legs of my tripod. She came out slowly, looking up at me. It felt as if she was looking for my confirmation that the coast was clear.

Strangely, I found myself whispering to her.

'Yes, it's safe to come out now,' I told her. 'The Sparrowhawk has gone.'

As if reassured by my words, and after one more glance up at me, she emerged from her safe haven and began to feed again.

I stayed with this Snow Bunting for a little while longer before leaving her to her beach. I had arrived to Annagassan that day in the hope of getting some nice images, and I had achieved that goal. However, I left that beach with so much more than nice images. I left feeling enriched by a truly remarkable life experience.

Such magical experiences were now, and still are, my new 'high' of birding.

Chapter 30 ➴

SO, WHAT'S MY FAVOURITE BIRD?

O ver the last ten years I have been working with national schools all over Ireland. I love working with school kids, sharing my love and passion for our precious birdlife with young birdwatchers of the future. It's funny, no matter what school or area I visit, one question is asked of me more than any other, and, I hasten to add, not just by school kids. That question is: 'What's your favourite bird in the world?'

At face value that seems like a very easy and straightforward question to answer but it's one that challenges me every time. There are over 10,000 species of bird in the world, of which I have experienced almost 3,000 on my travels. That's a lot of species to choose from.

I love waders and gulls, seabirds and birds of prey. Who sees a Snow Bunting and doesn't instantly fall in love with them? The subtle beauty and elegance of a Buff-breasted Sandpiper is also very special.

However, if I was to start narrowing it down, I have to say that Swifts have always been among my favourites. Their aerial abilities have always impressed me. Of course, the whole family of hummingbirds are definitely high on the list. I love everything about hummingbirds. Yes, 'hummers' are right up there amongst my favourites.

However, when I sit down and really think about it, there is one species that wins every time – the Swallow. Let me be taxonomically correct: by 'the Swallow', I mean the European Barn Swallow. In Ireland we have only one species of breeding swallow, so the name European Barn Swallow is simply shortened to 'Swallow'.

So, why Swallows?

There are so many reasons why Swallows are my favourite birds. They are such a part of Ireland in summer. Perhaps their arrival in spring stirs something primal within me – the caveman genes telling me that I have survived another winter? Perhaps it's their song. I know this is not exactly a scientific thing to say (I am not a scientist so I can say these things), but a Swallow's song always sounds joyful. It's as if they are happy to have made it home. Perhaps it is the epic journeys that Swallows make to reach home that attract me to them?

Irish Swallows leave Ireland in autumn and migrate across Europe, over the Mediterranean Sea and into North Africa. They then cross the Sahara Desert, travel over the equator and spend the winter in South Africa, 10,000 km from home. I have done that flight several times; it still takes an Airbus A380 almost thirteen hours to make that journey. It is a long way. Think about that. These birds weigh as little as 20 g (about half the weight of a bag of crisps) and are no bigger than your hand. Yet, they are capable of undertaking such migrations.

However, what they do in spring is even more remarkable. Not only do they return home to Ireland, but pairs, having gone their separate ways in autumn, will meet up again in the exact same barn or shed where they nested the previous summer, and raise another family. The male offspring from previous broods also return to their birthplace to nest, while the females tend to return to an area within a few kilometres.

If you live on a farm that has been in your family for many generations, you might enjoy this thought: the Swallows you will see this summer around your farmyard are direct descendants of Swallows that nested there when your father was a boy, or your grandfather was a boy and so on. The lineages of both species (you and them) share a direct connection to that farm.

As a child, I found it hard to imagine what Swallows experienced on their great journeys. I have since been fortunate to experience first-hand insights into what they endure. In spring, I have seen Swallows arriving into Majorca so exhausted that they were falling from the sky. They had faced into strong head winds crossing the Mediterranean from North Africa. We spent hours lifting these exhausted Swallows off the roads to save them from being killed by cars.

That any Swallow safely returns to Ireland in spring is a miracle of nature. I often think that each and every Swallow arriving home to Ireland deserves to be met by a full orchestra in celebration of their achievement.

In April 2011, I watched a Swallow come in off the sea in Co. Wexford. He (it was a splendid male) flew around and landed on a fence right in front of me. He had long tail streamers and a glistening plumage. He was obviously tired. He rested and preened. As I sat watching him, I found myself contemplating the minor worries and issues that had been preoccupying my mind over the previous twelve months. They were nothing to what this bird had endured during its phenomenal journey from one end of the globe to the other. I was sure he had faced many dangers and life-threatening moments that would make my issues pale in comparison.

This bird lived in the moment. He did not think about the past. He did not worry about the future. He had used all of his resources and wisdom to reach this far on his life's journey. He had no reason to reflect on the past or the future. This Swallow was living in the constantly shifting state of now.

I watched this wonderful bird for over an hour. The Swallow ignored me.

I left Wexford that day a little wiser.

Perhaps another reason I have a soft spot for Swallows is that these birds were the subject of the first conversation I had with a lovely person called Hazel, a close friend of a girl who was attending one of my bird courses. It was a phone conversation, and we discussed her concerns about her nesting Swallows in her shed. Little did I think when I spoke to her by phone that day that, in time, 'her Swallows in her shed' would become 'our Swallows in our shed'. Life does throw up surprises along the way.

That telephone conversation sparked a friendship that eventually led to me moving to Co. Wicklow, where Hazel and I now live. Here, in recent years, I've enjoyed watching Swallows nesting in our sheds around our home. I have watched them grow from helpless, 'baldy' little chicks, to young demanding Swallows almost too big to sit in the nest. I have enjoyed watching them make their first forays into the world, learn how to catch insects and master the skills of flight – skills they will need to get them safely to their wintering grounds in South Africa. It is estimated that more than 90 per cent of young Swallows die in their first year.

In autumn, I have watched huge gatherings of Swallows in Co. Wexford as they fed before flying out to sea to commence the first leg of their southern flight. Many of these birds were young birds, perhaps no more than six or eight weeks old. I have also watched flocks of Swallows along the very northern edge of the Sahara Desert before they braved the vastness of the rolling sands that lay ahead. Watching them muster up the courage to finally face the desert and then disappear into the shimmering horizon is unforgettable. And when I see a Swallow in South Africa, it is hard not to wonder if that bird, now flying across a lake where exotic Flamingos and Black-winged Stilts are feeding, is one of the chicks that were born in our shed.

My earliest recollection of seeing Swallows was on a sunny day in the Bots. I was a very young boy and Da was holding me by the hand as he showed me Swallows swooping low to drink from the ponds near the river. He was telling me about the great journeys the birds had made. He told me how they had come 'all the way from Africa and had crossed the desert to come home to Ireland'.

I was in awe. My young mind was enchanted by the journeys of these birds that were now flying around in front of me. I was filled with wonder at their coming home every summer to Ireland. That was almost fifty years ago. Yet, I can still remember that moment and those Swallows as if it was yesterday.

I believe it was hearing the story of the incredible journeys of Swallows that captured my imagination and helped to spark that sense of wonder in the world around me as a child.

It was a sense of wonder heightened by the encouragement of two very enlightened and extraordinary people, Ann and Tom Dempsey, my parents.

It is a sense of wonder I still carry.

It is a sense of wonder I still cherish.

EPILOGUE: HOME

One morning in August 2014, I sat by the open patio doors looking out into the garden of our home. Despite it being summer, it was a grey, muggy day. Our garden is a 'natural garden', with lots of trees that we have planted over the last three years mingling with the tall grass and wild flowers.

Grasshoppers were calling and a Red Admiral was feeding from some flowers close to the house. In the back of the garden are stands of taller, mature trees planted many years earlier. We have huge old Oak and Sycamore trees as well as lots of tall Ash trees. It is a little piece of natural paradise. It was from this stand of Ash trees that I heard something I never expected to hear in my own garden – the call of a Great Spotted Woodpecker. Imagine hearing a woodpecker in your own garden!

Great Spotted Woodpeckers are only recent colonists (or, more correctly, re-colonists) to Ireland. They had been absent for hundreds of years and only in the past ten years have they established themselves again as Irish breeding birds. I never thought I'd see the day when I could go into a Wicklow woodland and expect to see or hear a woodpecker. Now, here was one in our garden. How lucky was I to live here? How lucky was I to call this place home?

My thoughts were interrupted by the mewing calls of a young Buzzard as it soared over the house. It was one of three young birds that the local Buzzard pair had reared this summer. The previous day, I had watched a large female Peregrine Falcon pass her prey onto one of her brood, to great vocal excitement. That happened right over the trees at the end of the garden. Sometimes I have to pinch myself.

This summer, all our hard work planting trees had paid off. The number of birds that brought their young into the garden has taken me by surprise. Besides the usual suspects, we have played host to families of Redpolls, Siskins, Bullfinches and Linnets. Our dawn chorus this summer included Whitethroat, Blackcap, Chiffchaff and Willow Warbler, while our Yellowhammer was still in song up to early August.

We have also watched a pair of House Sparrows successfully fledge five chicks (in two different clutches), while a pair of House Martins has nested on the side of the house. They were late breeders, fledging three chicks during the final weeks of September. Each autumn we enjoy watching the aerial skills of large feeding flocks of House Martins and Swallows over the garden. The flocks then seem to drift off towards the coast, possibly to roost in the reed beds of Broad Lough.

Throughout September, there is a real feeling of autumn about our surroundings. Willow Warblers and Chiffchaffs constantly pass through. Our garden lies at the end of a shallow valley, and the amphitheatre of trees we have created in our garden acts as an ideal feeding place for migrants. They seem to particularly enjoy flitting around 'Tom's Copse', a small collection of Birches and Wild Cherry trees planted in Da's name. We placed some of his ashes into the holes of each of these trees as we planted them. They seem a fitting tribute to Da.

One of my daily autumn morning walks around the garden yielded fourteen Spotted Flycatchers. Imagine seeing fourteen migrating Spotted Flycatchers in your garden! They even granted me the pleasure of posing for some shots. And, as if this wasn't enough, during the last few days of October, I found a delightful and very vocal Siberian Chiffchaff feeding in Tom's Copse. This little 'eastern' warbler, which comes from Siberia, should have been making its way towards India to spend the winter there. Yet, here it was feeding on late-flying insects in our garden. Again, I sometimes need to pinch myself.

In 2013, we put a small pond in the middle of the wildflower meadow (for the want of a better word) and were delighted to find frogspawn this spring. That excitement was nothing when compared to the discovery of a Smooth Newt in the bottom of the pond a few weeks later. In summer, Dragonflies and Damselflies hover over the slightly green water. Swallows even drink from the pond while, late in the evening, Pipistrelle and Long-eared Bats hunt insects over it (we use a bat detector to identify them). We have seen Pygmy Shrews and Field Mice in the garden, and Rabbits graze on the lawn each morning. We even had a resident family of Hedgehogs pay us midnight visits for food at our back door in summer. A Fox also takes a nightly stroll through the garden and deer come to check out our tasty trees when they think nobody is looking.

In winter, our garden is alive with birds. We have feeders up everywhere and flocks of birds move in dizzying rotations. On one day, I counted over 100 Goldfinches feeding together on the feeders; add to these all the Siskins, Redpolls, Greenfinches, Chaffinches, tits, sparrows and the odd Collared Dove and Yellowhammer. Jays are also daily winter visitors. We have also enjoyed the company of Bramblings, Skylarks, Meadow Pipits and Reed Buntings. We've even had Red Kites flying over the house … imagine that!

Seriously, how lucky am I to call this home?

Thinking about this, I ask myself, Is it the garden and the wildlife it attracts that makes this place feel like home?

What makes a house a home? What does 'home' even mean?

I turn to Da's trusted old *Chambers Everyday Dictionary* that always lies close to my desk. This defines home as the following: 'habitual abode, residence of one's family, the scene of domestic life, one's own country …'

While these are all very clear-cut and straightforward definitions of the word home, they somehow don't quite catch the real meaning of home for me. There is the old saying that 'home is where the heart is' – that feels closer to the truth.

I wrote in the very first pages of this book that I have always referred to the house where my parents lived as home. I think most of us do. If we are lucky, home is where we grew up, felt safe and had our childhood dreams and ambitions. As adults, we move on.

We live in new places and, for many, those places become home. For others, those places never quite feel like home.

I am a northside Dub. I was born and raised in Finglas. I am a man who, up to the age of fifty, had never lived anywhere but on the northside of Dublin. 'You can take the man out of the northside but you can never take the northside out of the man'.

Yet, I now consider Co. Wicklow, where Hazel and I now live, to be my home. We have three cats (we had four, but Frigger has recently passed on and is buried in Tom's Copse). We have a beautiful spaniel called Suzie. We have a garden full of wildflowers, trees, insects, mammals and birds. We have a lovely house.

So is it all of the above that makes this house a home?

Yes, it is all that, but it is also so much more.

For me, this is home because I can truly be myself here. This is home because I am accepted for who I am here. This is home because we know laughter and love here. This is home because I have found peace here. This is home because I feel complete here.

'Here' just feels right. It's like all the pieces of the jigsaw are in place.

Like the returning Swallows, I feel I have reached my sanctuary for this part of life's voyage.

I am home.